Praise for
SPIRIT OF NATURE

"A naturalist's delight, this gem of a book is lavishly illustrated and written with obvious love and respect for the natural world. Highly recommended for the budding or experienced naturalist or indeed anyone interested in the natural world."

RICHARD P. READING, *PhD, Yale University,*
VP of Science and Conservation at Butterfly Pavilion

"Brian is a keen observer of the plants and animals around him. This well-written book is interesting and full of insight. It appeals to a wide audience. Brian has a strong conservation ethic. I highly recommend reading this book."

SUSAN CLARK, *Professor Adjunct, Yale University,*
School of the Environment

SPIRIT OF NATURE

in Northern New Mexico

SPIRIT OF NATURE

in Northern New Mexico

FIELD NOTES
on Natural History of Rio Mora National Wildlife Refuge

BRG
SCIENTIFIC

BRIAN MILLER, PhD

Spirit of Nature in Northern New Mexico:
Field Notes on Natural History of Rio Mora National Wildlife Refuge
Copyright © 2025 by Brian Miller, PhD

All rights reserved. No portion of this book may be reproduced in any form without permission from the publisher, BRG Scientific, except as permitted by U.S. copyright law.

For rights and permissions, please contact: BRG.Scientific@gmail.com or
BRG Scientific
3025 Ontario Rd NW #407
Washington, District of Columbia 20009

www.BRGScientific.com

ISBN: 979-8-9890369-3-6 (Hard Cover)
ISBN: 979-8-9890369-4-3 (Paperback)
ISBN: 979-8-9890369-5-0 (eBook)

Library of Congress Control Number: 2024927639

Editor: Melissa Stevens
www.PurpleNinjaEditorial.com

Cover Design and Interior Formatting: Becky's Graphic Design,® LLC
www.BeckysGraphicDesign.com

Cover photo of bison bull by Anabella Miller
Cover photo of peregrine falcon sourced from Adobe Stock

To Carina, Mary, and Anabella
for enriching my life

Contents

Introduction ... 1
   Why Me? ... 1
   Who Is This Book For? ... 2
   Why This Region? ... 3
   How Is This Book Organized? ... 4

PART I   Rio Mora National Wildlife Refuge: The Start of a Protected Area and Its Flora and Fauna

1. Formation of the Rio Mora National Wildlife Refuge ... 9
   Beginnings ... 9
   Our Arrival ... 11
   Trouble with Transition ... 12
   Becoming a Refuge ... 13
   Enter the Denver Zoo ... 15
   What We Accomplished with Our Programs ... 17
   Securing the Future of Rio Mora NWR ... 22
2. Landscape Around Rio Mora National Wildlife Refuge ... 27

PART II   Field Notes on the Natural History of Rio Mora National Wildlife Refuge

3. January and February ... 39
   January 1 to February 28/29 ... 39
4. March ... 51
   March 1 to March 15 ... 51
   March 16 to March 31 ... 58
5. April ... 69
   April 1 to April 10 ... 69
   April 11 to April 20 ... 73
   April 21 to April 30 ... 80

6. May ... 91
   May 1 to May 10 ... 91
   May 11 to May 20 ... 96
   May 21 to May 31 ... 101
7. June ... 105
   June 1 to June 10 ... 105
   June 11 to 20 ... 112
   June 21 to June 30 ... 116
8. July ... 123
   July 1 to July 10 ... 123
   July 11 to July 20 ... 128
   July 21 to July 31 ... 131
9. August ... 133
   August 1 to August 10 ... 133
   August 11 to August 20 ... 141
   August 21 to August 31 ... 143
10. September ... 147
    September 1 to September 10 ... 147
    September 11 to September 20 ... 150
    September 21 to September 30 ... 154
11. October ... 157
    October 1 to October 15 ... 157
    October 16 to October 31 ... 160
12. November and December ... 163
    November 1 to December 31 ... 163

Concluding Thoughts ... 169
Acknowledgments ... 173
APPENDIX A: List of Vertebrate Species ... 175
Endnotes ... 191
Reference Material ... 199
About the Author ... 211
About BRG Scientific ... 213

"There are some who can live without wild things, and some who cannot."

—Aldo Leopold, *A Sand County Almanac*[1]

Introduction

Why Me?

Throughout my forty-year career, I avoided being deskbound. Agencies are bureaucratic by nature, and I never had a desire to run laps around the academic tenure track. I wanted to stay outdoors as much as possible. Nature holds a lifelong fascination for me, and since 2005, I have been part of northern New Mexico. I love watching the change in seasons and the way nature adapts to them. I look forward to learning more each year. Every answer opens more questions: When does a certain species of flower bloom? When does a given species of migratory bird arrive in the spring to build nests and lay eggs? Can changing conditions affect that timing? How do organisms adapt strategies for survival, particularly when times are harsh? Which songs belong to which birds? Which mammal tracks should I follow in fresh snow? Which fragrant scents can we expect in different parts of spring and summer? What new surprises are waiting at each new excursion? How did nature rebound from an unseasonal, extreme event?

I remember one such rebound vividly. We sometimes have April snow that falls on blooming pasqueflowers (*Pulsatilla patens*). In 2005, we had 26 inches (66 centimeters) of snow in late March. The snow was spread over three days of constant flurries. Before that, it had been warm, and ground squirrels were chirping while birds sang. Then one morning, everything was silent. That afternoon, the snow started. Three mornings later, it was still snowing, but the birds still started singing after sunrise. About two hours later, the snow stopped. How did those birds know when the snow would start and stop? While it is not uncommon for organisms to sense coming weather changes, all the *cannots* of the Leopold quote: "There are some who can live without wild things, and some who cannot" ponder these things.

Because I have spent a good portion of my research career outside, I have learned that by noting the timing of concurrent events, an observer of nature can know that by seeing one event, a person should be able to predict and search for another event happening at the same time. Of course, Native Americans have done this for centuries. This book examines and expands on the timing of events for a given species to describe how it adapts to the changes in seasons. For example, how does hibernation help bears get through a winter when much of their diet is unavailable, or how does delayed implantation benefit badger reproduction?

Who Is This Book For?

I've written this book for the amateur naturalist or nature lover who doesn't mind muddy shoes or sopping wet socks with seeds sticking in them. I've provided some pictures and many hand-drawn sketches by myself and my daughters of several of the species mentioned. If you're interested in going deeper into field exploration, I recommend that you purchase some guidebooks with detailed images of the species that you seek. A handy pocket guidebook for this region is the *National Audubon Society Field Guide to the Southwestern States*. It covers both the flora and fauna of New Mexico.

I've provided scientific names at the first mention of a species. In some cases, common names can differ in different regions, so the scientific name will justify the use of a common name after the first mention. For example, the common name raccoon can also be *mapache* because of Spanish influence in the southwestern United States. The binomial system using Latin for genus and species, however, doesn't change across regions.

There is also a word for the study of repeating patterns in nature: phenology. When many people think of seasons, they think of four because that is what the calendar tells us. When following changes in nature, however, there are more like twenty-five seasons. That's why I've divided part two of this book into (mostly) ten-day segments.

Climate is an important factor in phenology, and shifts in weather can change the early or late actions of Nature's clock. Climate concerns long-term patterns whereas weather signifies short-term events. Even if weather varies phenology, the associations will still occur at the same time. You will see the same group of events at the same time annually, whether it is (say) in late April or early May. A late freeze may slow events temporarily. Cottonwood buds will simply arrest development in a late freeze until more agreeable weather lets development resume. That said, climate change may break some associations. Long-term changes in climate and rain patterns may favor some species at the expense of others.

Why This Region?

The area that inspired me to write this book is near the terminus of the Santa Fe Trail on the high plains of northeastern New Mexico. My family's life has revolved around this area since 2005. We came to manage the Wind River Ranch because the owner, Eugene Thaw, wanted to see his land serve as a societal and conservation benefit. That ranch became the Rio Mora National Wildlife Refuge in 2012, and the pattern of nature here reflects the sequence of events elsewhere on the high plains of northeastern New Mexico.

Altitude and latitude affect phenology and habitat. For every 328-foot (100 meters) increase in altitude, the temperature decreases by 1.8 degrees Fahrenheit (1 degree Celsius). Latitude is the distance of a given location from the equator, and for each one degree moving to the north, temperature declines 1.2 degrees Fahrenheit (0.7 degrees Celsius). Because the planet is round, the angle of sunlight hitting the Earth changes according to where you stand at the moment. At the equator, solar energy strikes the Earth at a nearly direct angle, and the shorter distance sunlight travels through the atmosphere means that less heat is lost. As you get nearer to the poles, the angle of sunlight grows more acute, and solar energy must travel a longer distance through the atmosphere to arrive. Thus, more heat is lost as it travels through the atmosphere than is the case at the equator. The direction of a slope can also create microclimates. A south-facing slope will receive more solar energy than a north-facing one.

The Rio Mora National Wildlife Refuge is located on the high plains near the foothills of the Southern Rockies. The 4,225-acre (1,710 hectares) refuge ranges between 6,700 feet and 7,000 feet (2,042 to 2,134 meters) with the Mora River running through it. The Mora River starts at 10,000 feet (3,048 meters) of altitude on Osha Mountain in the Sangre de Cristo range. It then flows eastward to join the Canadian River at approximately 4,500 feet (1,372 meters) of altitude. Whether you are farther west or farther east along the Mora River will affect phenology sequences, but the heart of this book is the Rio Mora National Wildlife Refuge, which is midway between the start and finish of the Mora River.

How Is This Book Organized?

I start with a bit of history about how Wind River Ranch became Rio Mora National Wildlife Refuge, then follow that with a bit of ecological history of the region. After those two chapters comes the phenology and natural history of the area.

Writing this book was a joy. It is a celebration of nature in a place that I love. Writing it was a cathartic process in an otherwise chaotic time of Covid and political partisanship. I hope the book inspires readers to get outside and explore wild places. Experiencing nature can be beneficial for one's mental health. Given the lack of action on climate change, it is best to do it sooner than later.

PART I

Rio Mora National Wildlife Refuge: The Start of a Protected Area and Its Flora and Fauna

CHAPTER 1

Formation of the Rio Mora National Wildlife Refuge

Beginnings

The landscape of the Wind River Ranch in northern New Mexico—within the transition zone of the Sangre de Cristo Mountains to the west and the Great Plains to the east—is both valuable to nature and ecologically important.

Eugene and Clare Thaw, the most recent owners of the ranch, were from New York, but they traveled to New Mexico often because Eugene was an assessor for Georgia O'Keefe's estate. When Eugene retired from his New York art dealership, the Thaws moved to Santa Fe, and they bought the Wind River Ranch in Mora County as a recreational ranch to complement their Santa Fe home. Eugene retired because he felt the rapidly increasing wealth of this generation was turning interest in art from a pursuit of beauty into merely a financial asset.

Born in 1927, Eugene Victor Thaw was named after Eugene Victor Debs, who had died the year before. For those who do not recognize that name, Eugene Debs was a socialist, labor organizer, and peace activist who had been nominated for the Nobel Peace Prize. Debs advocated for many things that we take for granted today, such as the eight-hour workday and the end of child labor.

The Thaws didn't come from money. Eugene Thaw's father was a heating contractor and his mother a schoolteacher. Eugene didn't have the money or social connections to curate a museum, so he opened a small gallery in a rented room to sell art. He frequented small auction houses and antique stores to get stock for his gallery. When his success as an art dealer increased, he hired Clare to be his gallery assistant. They were soon married. Eugene became one of the most respected art dealers in the field as well as a global expert on old European Master art.[2]

Figure 1.1. Clare and Eugene Thaw
Photo: Thaw Charitable Trust

The Thaws were exceptionally generous. In 1981, they started the Thaw Charitable Trust as a philanthropic organization. The initial funding came from selling a single Van Gogh painting for $40 million. The main interests of the trust were art, the environment, and animal welfare.

Their generosity helped numerous projects. Clare died in 2017 on her ninety-third birthday, and Eugene succumbed six months later. They had been married for sixty-three years.

In 2004, Eugene decided that he wanted the ranch to serve a societal benefit, so he offered the land to the Wildlife Conservation Society (WCS) of New York along with $10,000 a month for ten years to cover expenses. The WCS is a consortium of the zoos and aquariums in the five boroughs of New York City. It aims to conserve the world's large wild areas.

I was employed as a conservation biologist at the Denver Zoo at that time, and WCS staff asked me to come to the ranch and manage it as their employee. I wrote a document for WCS about what I thought we could do at Wind River Ranch and shared it with Eugene. He liked the ideas I proposed on conservation of native species and ecosystems, conservation research, restoration, and education. However, the board of directors at WCS decided to reject Eugene's offer near the end of 2004. The reason is unclear. Eugene wanted to see the ideas implemented, so when WCS rejected his offer, he started the Wind River Ranch Foundation on his own. He thought that if the ranch was functioning as a societal benefit, surely WCS would reconsider in another year or two. I started in January 2005 as an employee of Wind River Ranch Foundation, and we began the programs that year. Thaw Charitable Trust covered two salaries and operating expenses, and those expenses were handled by Sherry Thompson, who was the executive director of Thaw Charitable Trust in Santa Fe. The programs were all funded by grant-writing.

Our Arrival

I arrived at the ranch in January 2005. My wife, Carina, was a teacher's aide in a Denver public primary school where both of our daughters attended. Carina initially had reservations about leaving Denver to live in a double-wide prefab home at 7,000 feet (2,133 meters) on the high plains of New Mexico. My wife and daughters stayed in Denver to finish the school year while I worked at the ranch. They joined me in May.

From 1999 through 2004, while at the Denver Zoo, we spent five summers at the University of Wyoming/National Park Service Research

Station located on Jackson Lake in Grand Teton National Park. We lived in a one-room cabin at the research station while examining how re-colonizing wolves affected coyotes and how that effect trickled down to smaller animals. So, my daughters, Mary and Anabella, were exposed to nature from a very early age. They were excited about living on the ranch, and Carina came to like it too. Carina became a teacher's aide in the bilingual program at a primary school in Las Vegas, New Mexico. Spanish was her first language, and we met and married while I was living in Mexico. Essentially, I had to leave the country to find someone who would put up with me.

In the fall of 2005, our two daughters entered second and fourth grade in Las Vegas, New Mexico. In our first fall at the ranch, the sunflowers bloomed all along our lane to the house, and my daughters would race on the half-mile (1 kilometer) lane pretending to be in the Olympics with the sunflowers as fans cheering them on. Both went on to run track in high school, and Anabella continued on to run at New Mexico Highlands University. Another fond memory was hearing them around a corner of the house saying, "This is the best mud day yet."

Trouble with Transition

Eugene wanted to donate the ranch to an NGO because he was averse to government bureaucracy. He offered it to several organizations between 2007 and 2010, but the 2008 economic crash made NGOs reluctant to accept the land without a full endowment. While he was willing to give a transitional stipend, he did not want to give a full endowment that would cover all future costs. He wanted the NGO to recognize the value of the land to conservation and thus invest some of their funds into it. He was adamant that the land should belong to an NGO and not the government. Indeed, the director of the Smithsonian/National Zoo Conservation and Research Center, Dr. Chris Wemmer, was interested in adding the land to their conservation interests. When I told Eugene, our conversation went something like this.

Eugene: "No, Smithsonian is government."
Me: "It is only half government; the other half is private."

Eugene: "Half government is too much."

In 2007, we started a partnership with the Jicarilla Apache to graze their bison. The elders in their Cultural Affairs Office wanted to collect plants from this area because their lands used to extend from Dulce in north central New Mexico to the Canadian River in eastern New Mexico. I invited them to come to Wind River Ranch for those plants. While here, they said they had received ten bison from South Dakota but didn't have permission from the tribal council to graze those animals on Reservation land. I suggested that they graze the bison at the ranch until they could get permission.

At that time, we also reintroduced prairie dogs (*Cynomys* spp.) to Wind River Ranch. Mora County had an ordinance prohibiting people from bringing prairie dogs across the county line. The County Commission overturned that ordinance so that we could reintroduce prairie dogs. That may be the only time a western county has overturned a rule against prairie dogs.

In 2008, the Inter-Tribal Bison Cooperative gave the Jicarilla thirty-five more bison. The Inter-Tribal Bison Cooperative was a consortium of fifty-five tribes reestablishing bison on their lands. Because we were taking care of the Jicarilla Apache bison, they gave some of their animals to Wind River Ranch Foundation, so we were splitting the herd. In 2009, the tribal government changed, and they received permission to graze their bison on Jicarilla Apache lands. Instead of removing half the bison to Dulce, which would break the herd dynamics, I raised grant money to buy the Jicarilla Apache bison, and they used the money to start a herd on their lands. The bison were now owned by Wind River Ranch Foundation. Bison and prairie dogs formed the grasslands that are ranched today, but for their reward, they were shot and poisoned. They are key species for the prairie.

Becoming a Refuge

It was early 2010, and no NGO wanted the land without a full endowment. The rejections of Eugene's generosity wore him down, and he decided to sell the ranch on the market. The day after he announced it, a realtor showed up with a potential buyer, and that was a very dark day for me. We had started programs with regional middle and high schools from nineteen school districts, had ongoing ecological restoration projects, bison and prairie dogs for grassland restoration, graduate projects from New Mexico Highlands University, and had become a place where agencies and NGOs could come to discuss conservation issues. That work would have been lost if the land became a cattle ranch.

The next day, I asked if the USFWS would buy the ranch and make it a National Wildlife Refuge. Rob Larañaga, manager of the Las Vegas National Wildlife Refuge/Northern New Mexico Refuge Complex, stepped forward and said the USFWS would purchase the land to form a refuge. Rob and Dr. Benjamin Tuggle, Region 2 Director of USFWS, said they would maintain the programs and partnerships started by Wind River Ranch Foundation, and they would provide a budget for staff and operating expenses. Because USFWS would maintain the original vision and fund the operation, Eugene decided to donate the land to USFWS. Importantly to us, keeping the mission of the Wind River Ranch Foundation meant maintaining the refuge for native species only, continuing education programs, conservation research, and ecological restoration efforts.

The catch was that some USFWS folks in the regional office in Albuquerque were not fond of keeping the Wind River Ranch Foundation bison after the land entered the refuge system. Teresa Gray, who was an employee of Wind River Ranch Foundation, and I worked on a deal with Gabe Montoya and Phil Viarreal to switch ownership of the bison to the Pueblo of Pojoaque tribe. Officially, the sale was for an undisclosed sum, but it was fifty-five bison for fifty-five buffalo nickels. The sale was complete in 2010. Having tribal bison on the land as a partnership meant it would be harder for USFWS to remove the animals. Our hope was that the model of federal lands being a partner with tribes to graze tribal bison would inspire other federal lands to do the same.

Between 2010 and 2012, USFWS had to go through a process to accept the land. This involved public meetings. During the three public meetings, people in the audience advocated for us. The support largely came because the community had stakeholders in the programs, particularly in education efforts for schools of the region.

Enter the Denver Zoo

Unfortunately between 2010 and 2012, the USFWS suffered budget cuts. At the time, Region 2 of USFWS had sixty open positions and lacked funding to fill them. So, Eugene donated $1,715,000 to the Denver Zoo to pay salaries and manage the refuge for USFWS. The employees of Wind River Ranch became Denver Zoo employees. He also hoped that his donation would be used as a seed to build an endowment and to produce a plan for long-term financial viability. Programs were still dependent on our grant-writing, but Eugene's donation to the Denver Zoo indirectly benefited the USFWS and meant that programs and staff could expand in number and efficiency. The staff grew to four, plus temporary interns.

On September 27, 2012, the Wind River Ranch officially became the Rio Mora National Wildlife Refuge (Rio Mora NWR). Rio Mora NWR was the 560th National Wildlife Refuge in the USFWS system, and it became part of the Northern New Mexico Wildlife Refuge Complex, joining the Maxwell and Las Vegas (NM) National Wildlife Refuges. Rio Mora NWR also became the core of a 952,000-acre (385,261 hectares) USFWS Conservation Area in the Mora River Watershed. The Conservation Area designation meant it would be easier to grant easements and work with surrounding landowners. There are only two such conservation areas around national wildlife refuges in the United States. We had already cooperated with about 300,000 acres (121,406 hectares) of private land in the area as Wind River Ranch. Our value was in developing conservation techniques that could be exported outside of the refuge boundaries. Those restoration techniques are still being used by neighboring ranches today.

When the Wind River Ranch became Rio Mora National Wildlife Refuge, the local newspaper, *The Las Vegas Optic*, listed it as one of the top ten events in 2012. It is not often that a rural, western newspaper thinks that the feds getting more land is a good thing, but we had excellent community support.

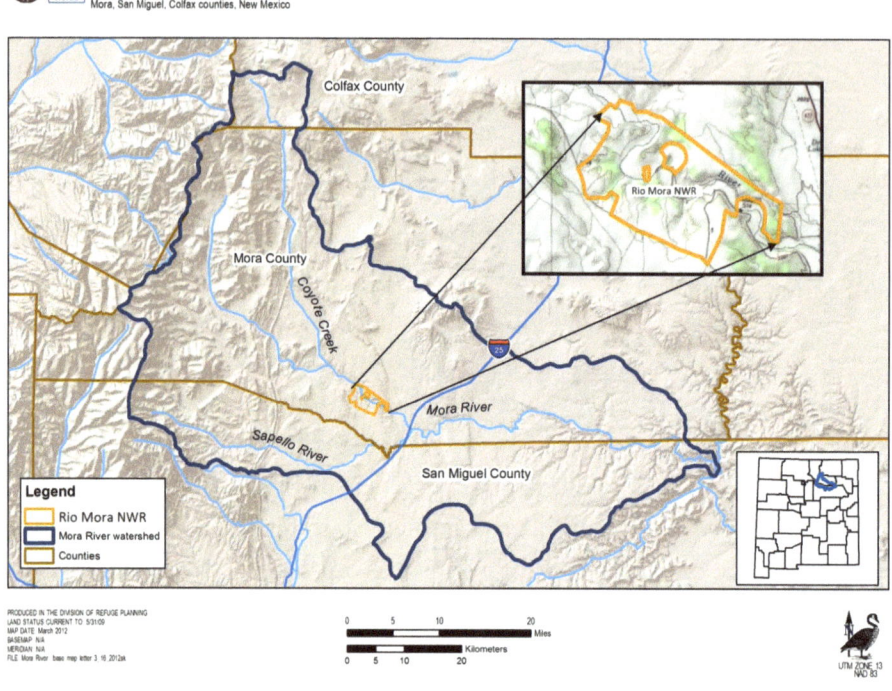

Figure 1.2. Map of the 952,000-acre (385,261 hectares) conservation area around the 4,225-acre (1,710 hectares) Rio Mora NWR (noted by the black dot)

Map courtesy of the United States Fish and Wildlife Service

I retired in 2014, two years after the land became Rio Mora NWR, knowing that Rio Mora NWR embodied our dream. The programs were run by the capable hands of Luis Ramírez and Shantini Ramikrishnan from 2014 until February 2019, then by Shantini until USFWS placed employees on the refuge in 2020. I became a daily volunteer. I helped when I could and stayed out of the way when I couldn't.

The management of Rio Mora became a formal partnership among USFWS, Denver Zoo, Pueblo of Pojoaque tribe, and New Mexico Highlands University (NMHU), an Hispanic-serving institution. Each of the partners contributed between $20,000 and $40,000 a year in cash directly to Rio Mora NWR from their organizational budgets or through in-kind services. Denver Zoo managed the Thaw donation, which was essential because salaries and operating expenses amounted to about $325,000 a year. Because there was no federal money, there were no USFWS staff on the refuge. We did the work of managing the land and running the programs. There was a strong spirit of cooperation among the partners.

What We Accomplished with Our Programs

From 2005 until the end of 2019, our science emphasis was on applied conservation knowledge that could be used on lands outside of refuge boundaries. We had good relations with about 300,000 acres (121,406 hectares) of neighbors and with the northern Pueblo tribes. Research with direct conservation value—restoring ecosystem function and protecting native species—were the priorities. Our efforts at restoring grasslands and wetlands helped buffer the refuge from a changing climate. As of January 2020, we had hosted twenty-nine studies for an MS, two for a PhD, one post-doc, and twelve by working biologists. Since 2005, we had hosted sixty-seven internships, and 90 percent were filled by students of color. We hosted international interns from Mexico, Papua New Guinea, Cameroon, and Haiti. While these numbers are modest compared to older research centers, we always have had a small staff.

Rio Mora NWR had played a strong role in STEM education for New Mexico youth and university students. Thus, New Mexico Highlands University was a strong partner. Since 2005, over six thousand students from nineteen school districts, eleven tribes, and eleven universities have come to the refuge for classes about conservation and science and to participate in citizen science research. Much of this has been due to the hard work and creativity of Shantini Ramakrishnan, who began working at Rio Mora when the land was still Wind River Ranch.

Roughly 90 percent of the students coming to Rio Mora were New Mexico residents. This was important to us and to the local community. New Mexico was ranked fiftieth in preparing high school students for college and forty-ninth in education. Around two-thirds of a freshman class in Las Vegas graduate from high school after four years. San Miguel

County has a median household income of $29,000, and Mora County has a median household income of $21,000. Rural New Mexico remains a forgotten area, and most Rio Mora resources were spent locally.

Our partnership with the Pueblo of Pojoaque tribe was strong, and at the time it was unique in the National Wildlife Refuge system. Management of the National Bison Range, which is a National Wildlife Refuge, was returned to the Confederated Salish and Kootenai Tribes in 2020. The Pueblo of Pojoaque bison herd also serves other southwestern tribes. The herd provides ecological restoration on the land and cultural restoration for the tribe. We hosted tribal youth camps at Rio Mora NWR, which of course are led by the tribes.

In our restoration efforts, grade control rock dams in arroyos raised the water table, recharged springs, and restored degraded wetlands. They are called one-rock dams because they are only one rock high. Even though they are called dams, they don't stop the water. They slow the flow allowing water to soak into the soil. Wetlands and riparian areas represent less than 1 percent of the total area in the Southwest, but about 75 percent of vertebrate life depends on these areas for food, water, cover, and movement.[3] As of the end of 2019, we, along with our volunteers, had built over 450 grade control structures in arroyos and canyons. They work. Bill Zeedyk led these efforts, and he pioneered the techniques. Some of the arroyos on the refuge formed during the overgrazing in the 1880s.

Albuquerque Wildlife Federation was a key volunteer group for the construction. They built more structures on the refuge than any other group. They returned twice a year, each time with a large group of trained workers.[4]

Figure 1.3A. Silva Canyon before

Photo: Brian Miller

Figure 1.3B. Two layers of rock structures that covered bedrock by 3 feet (0.9 meters) of sediment and produced ample vegetation where there was once none

Photo: Brian Miller

We monitored changes in soils, moisture, vegetative recovery, and wildlife use in restored arroyos and canyons, and we compared those variables to arroyos not yet restored. Arroyos across sloping grasslands, which are typically formed by legacy ranch roads, disrupt the natural water flow across the landscape and deprive grasslands down-slope from the road of needed moisture.[5] One study on the refuge explored a technique to get water out of the old roads and back onto the down-slope grassland.

Controlling upland erosion was a key to nearly all restoration efforts on the refuge. Upland erosion degrades soils, reduces plant diversity, reduces plant densities and cover, lowers the water table, and increases sediment in the river.[6] Erosion happens when there is not enough plant cover to protect the soil surface from the sheer force of flowing water; erosion worsens as the slope increases in angle, as the slope lengthens, and as the soil becomes more fragile. Where a grade suddenly becomes steep, or soil hardness changes, run-off creates a head-cut, likened to a dry waterfall—at least until it rains. Head-cuts, in turn, increase the speed of run-off after rains, deepen the arroyo, and lengthen it. The erosion continues to the river, changing water quality, temperature, and oxygen content of the river. As the velocity of flow within an arroyo increases, the power of erosive force increases exponentially (doubling the velocity cubes the energy of water).

One of the most important features of soil is its ability to hold water; degraded soil loses that capacity. In the Southwest, rainfall is scarce, and climate change that increases drought will worsen the effect. Scarce rainfall is complicated by unintended water loss through evaporation. To return life to the land, soils must recover the ability to hold water. Slowing the flow of water allows it to soften the ground and increases the amount of infiltration into the soil. Water stored in the soil is less vulnerable to evaporation, it flows slowly downhill (subsurface) to feed seeps, wetlands, and springs, and it bolsters riparian habitat. Soil moisture boosts micro-organisms, such as fungal mycelium, that help transport water to plant roots. Restoring plants to a degraded area slows air movement, provides shade, reduces evaporation, stores carbon, slows water flow, holds soil in place, and provides ground litter that adds organic content to soil. This structural change provides habitat for a host of vertebrates and invertebrates.[7]

Another major effort was river restoration. The Mora River had incised and had been disconnected from its floodplain. That process lowered the water table and eliminated important riparian habitat. Stable rivers that meander across valley floors have functional floodplains, or areas that occur naturally along the stream where the river deposits water during flood events. When flood waters spread across the vegetated floodplain, they spread the energy of the river. By dissipating energy and slowing flows, floodplains reduce erosion of the bank and bed of a flooding river. Floods deposit rich sediments onto the floodplain, recharge water tables, create diverse habitats, and sustain communities of plants and animals. A river disconnected from its floodplain is a river deprived of its ecological function.[8]

Wind River Ranch Foundation, in cooperation with Quivira Coalition, Environmental Protection Agency, and USFWS Partners for Fish and Wildlife, used "induced meandering" to restore a 0.6-mile (1 kilometer) stretch of the Mora River in 2010. Years ago, that stretch of river had been straightened for farming. Induced Meandering tweaks the system to help water flows speed the natural process to recreate a meandering channel. When rivers are straightened, the floor of the river flattens and moves away from the pool, glide, riffle, run sequence that is important for fish habitat. Induced meandering methods also restore that sequence on the floor of the river. Structures are made of natural materials, such as boulders, cobbles, posts, tree trunks, and living materials gathered from the area.[9]

The need for ecological restoration in the region is urgent. The negative impact of over-exploitation is clearly visible. Erosion, lack of water, and declining health of grasslands are issues of concern in our region, and 80 percent of New Mexicans agree that restoration of such ecosystems is important.[10]

Securing the Future of Rio Mora NWR

When Dr. Benjamin Tuggle was the director of USFWS Region 2, he gave us tremendous support. Indeed, he wanted all interactions over making the Wind River Ranch into Rio Mora NWR handled over his desk, skipping over the Region 2 head of refuges. But the Trump administration moved him to an office in Washington, DC. Perhaps this was part of Trump's efforts to destabilize federal agencies. Craig Piper was the director at the Denver Zoo in 2012 at the time of the transfer, but he accepted a position to be the director of the Central Park Zoo of WCS in the fall of 2013. Craig was a strong advocate for the Denver Zoo to continue running the programs at Rio Mora NWR, as was the head of the conservation department, Dr. Richard Reading. In January 2014, the Denver Zoo named a new director, but she had no previous animal experience. She was hired because of her fundraising record at another NGO. With this appointment, emphasis at the Denver Zoo shifted away from conservation and toward entertainment. It also meant that the Denver Zoo was adopting a corporate philosophy. In 2016, Rich Reading left the Denver Zoo in frustration over that shift. We thus lost our big advocates at both the Denver Zoo and Region 2 of the USFWS.

By 2018, it was clear that the $1,715,000 donation from Eugene Thaw to the Denver Zoo would expire at the end of 2019. They treated the donation as an open checkbook. There was no endowment and no plan for long-term security from either the Denver Zoo or from USFWS. Worse, the bureaucracies of both did not seem disposed to form the plan. The future of Rio Mora NWR was uncertain.

In July 2018, Shantini and I met with Senator Tom Udall, and he said he would find funding for the Rio Mora NWR in the Senate Appropriations Committee. While this was outside the normal USFWS channels, I was very open with USFWS about our effort to work with Tom Udall to fund Rio Mora. He was the ranking head of the subcommittee for environmental appropriations. The Udall family had a long history of environmental awareness. His dad, Stewart Udall, was the Secretary of Interior during the 1960s, and I think he was the best person to ever sit in that chair. Senator Martin Heinrich helped, and he also had a firm grasp of environmental issues. Our representative in the House, Ben Ray Lujan, supported what we did, but at that time the House of Representatives did not have an Appropriations Committee.

During 2019, Senator Udall pushed hard on the USFWS bosses in Washington, DC, basically telling them that USFWS had received a gift of land, as well as a donation covering nearly all expenses, and yet they had put very little into Rio Mora NWR. Not only that, but we did the day-to-day work for them. They had no USFWS staff at the refuge. This embarrassed Region 2 of the USFWS. Both New Mexico senators listed funding for Rio Mora NWR as a New Mexico priority in the 2019 federal budget that passed both congressional chambers and was signed by the president.

Senator Udall secured money from the Senate Appropriations Committee for USFWS to give Rio Mora NWR a budget. The Rio Mora budget was an add-on to the regular Region 2 USFWS budget, so no other programs were hurt. The USFWS fiscal year extends from October 1 of one year until September 30 of the following year. In October 2019, with the new budget in place, Region 2 of the USFWS placed two employees—a refuge manager and a biologist—at Rio Mora and began covering all operating expenses. Thanks to Senator Udall, Rio Mora NWR would remain open when the Thaw donation expired, which it did in November 2019. It was that close to having no budget.

In October 2019, the Denver Zoo had assured the Pueblo of Pojoaque that the Zoo would stay in the partnership after the Thaw donation expired. In January 2020, however, the Denver Zoo announced that they were going to transition out of the partnership. This put the programs at risk once again. The Denver Zoo said their withdrawal would be gradual, but then in the spring of 2020, came Covid-19. The Denver Zoo laid off 25 percent of their staff, including 75 percent of their conservation department. Those layoffs came several days before the Paycheck Protection Program began in April 2020. The Paycheck Protection Program was meant to cover salaries of organizations for twenty-four weeks, and the loans could be forgiven if the money was used for salaries. I don't know why the Denver Zoo did not extend the Paycheck Protection Program to those who they instead terminated. Shantini was retained by the Denver

Zoo as the only employee at Rio Mora. She was to tie up loose ends for the Zoo's exit, but her future with the Denver Zoo did not appear secure. A Denver Zoo employee later told her that the zoo was planning to lay her off in the fall.

Other groups, however, saw the value of collaboration and were willing to add support. Within New Mexico Highlands University, the president, Dr. Sam Minner, and faculty member Joe Zebrowski pushed for a larger role in the partnership. In August 2020, New Mexico Highlands University stepped forward and hired Shantini to coordinate the restoration, research, and education as part of the university STEM efforts. Hands-on efforts in STEM classes greatly increase success of students and retention. Dr. Edward Martínez, a university vice president, was given the role of institutionalizing those programs. Edward grew up in Mora, so he knew the area intimately. He specialized in water quality as a professor. The partnership became USFWS, the Pueblo of Pojoaque tribe, and the New Mexico Highlands University. After fifteen years, the Rio Mora NWR finally seemed to have a budget for long-term stability.

Then Covid swept the nation, and activities on the refuge began to stop. The partnerships with the Pueblo of Pojoaque and New Mexico Highlands had been strong, productive, and enjoyable, but those partnerships soon deteriorated. Indeed, some of the USFWS employees in the Northern New Mexico Refuge Complex, of which Rio Mora NWR is a part, did not like the partnerships that existed at Rio Mora NWR. One person even criticized the partnerships openly in 2018/2019 partnership meetings. Her attitude was that it should be USFWS alone. Such self-centered thinking exists in nearly all organizations and agencies.

Even with the original support from Benjamin Tuggle, some in the regional office had shown early resistance to tribal bison on the refuge. At the ceremony to celebrate opening the Rio Mora NWR in 2012, one of the USFWS regional bosses from Albuquerque told me they were going to get rid of the bison. We pushed back. We had hoped that the partnership with Pueblo of Pojoaque might encourage other tribes and other federal lands to cooperate over tribal bison. We were right, and the model was working until 2020.

After 2020, however, tension between the tribe and USFWS increased as bureaucracy (denying or slow-walking requests) made it harder for the tribe to manage their bison. Perhaps some of that tension came from us going directly to Senator Udall for a budget when no other options were available. The US government has done many things over the last several hundred years to raise the level of indigenous distrust. Tribes lost their lands to Manifest Destiny, not to mention the government role in eliminating the bison, and we originally thought that having tribal bison on federal lands would in a small way increase access to former lands. USFWS also wanted to remove New Mexico Highlands University from the partnership, apparently ignoring the expertise present in the university. New Mexico Highlands University was unaware of this plan, but the Pueblo of Pojoaque tribe defended the university and that defense kept the university in the partnership.

Despite receiving free land, expenses for eight years, and then a budget with no effort on the part of USFWS, bureaucracy and Covid hindered the education and research efforts by New Mexico Highlands University on the refuge. Shantini continued the education, research, and restoration efforts, but mostly on private lands, like the neighboring Fort Union Ranch.

Although the model degenerated, the Biden Administration and Secretary of Interior Deb Haaland started a program for tribal co-management of federal lands.[11] A national tribal lawyer, Daniel Rey Bear, initiated a cooperative management agreement among USFWS, the Pueblo of Pojoaque tribe, and New Mexico Highlands University. The cooperative management agreement will last for ten years. That agreement elevated the status of the tribe and the university in decisions at Rio Mora NWR.

This cooperative agreement, combined with the retirement of a key figure in the USFWS regional office (the one who didn't want bison on the refuge), has changed the tone of the USFWS toward Rio Mora NWR. The atmosphere is friendlier, and the partners can express opinions about new strategies. Meetings are more cordial and conducted with more humility. The USFWS is installing and funding wildlife-friendly fencing around the perimeter of the refuge, and that is an expensive undertaking to benefit the bison. The fencing should make it harder for bison to leave the refuge while allowing the passage of elk, deer, and pronghorn. The cooperative agreement is only in its first year, but we are hopeful. The cooperative agreement reinforces the original goals of research, education, and ecological restoration from the old Wind River Ranch Foundation. We hope that this leads to more engagement with the local community, neighbors, and tribes. USFWS is increasing its education efforts on Rio Mora NWR.

Partnerships require members to try and understand different viewpoints among partners. Consensus must replace unilateral action. Partnerships have not been very common on USFWS refuge lands across the nation. Many refuges are managed as isolated islands within surrounding habitat. The saying "save the dirt" has been commonly used about refuges, but *The Theory of Island Biogeography* in 1967 was a major breakthrough that showed no single park or refuge is large enough to maintain viable populations of large species, particularly large carnivores.[12] It showed that the smaller the protected area, the more species are lost due to lack of habitat, edge effects, vulnerability to catastrophic events, and genetic or demographic disorders. Given that more than 90 percent of habitat around the globe is located outside of protected areas, partnerships that reconcile human use and conservation needs are essential. Individual refuges functioning as islands are too small to protect biodiversity. Partnerships are the future for large scale landscape conservation.

I think a lesson here is that nothing can be considered permanent. One can't stop paying attention. Changes at the top of organizations can change goals or at least the order of priority for goals. Collaborative management means that permanent professionals in the agency must support it. It can't just be supported by political appointees who change with administrations. That said, the collaborative management agreement shows that policy can be shifted to a more positive direction and that cooperation goes much farther than an individual effort. Conservation requires science, active citizens, and government action. Don't expect someone in power to make the right decision. *Make* them make the right decision.

CHAPTER 2

Landscape Around Rio Mora National Wildlife Refuge

THE 4,225-ACRE (1,710 HECTARES) Rio Mora National Wildlife Refuge sits at 6,700 to 7,000 feet of altitude (2,042 to 2,134 meters) in northern New Mexico. The main habitat types are shortgrass prairie, riparian, ephemeral natural catchments, and perennial seeps/springs/marsh wetlands, piñon/juniper/oak woodlands (*Pinus edulis-Juniperus* spp.-*Quercus* spp.), and ponderosa pine (*Pinus ponderosa*) forests.[13] The refuge includes both sides of the Mora River, which is a sub-basin in the Canadian Watershed. The main tributaries feeding the Mora River are the Sapello River and Coyote Creek. The river covers a distance of 116 miles (187 kilometers) from Osha Mountain to the Canadian River near the tri-county border of Mora, Harding, and San Miguel Counties.

The Audubon Society designated the Wind River Ranch as an Important Bird Area in 2008. Important Bird Areas are sites that provide essential breeding, migrating, or wintering habitat for birds. They support one or more high-priority species, large concentrations of birds, exceptional habitat, and/or have substantial research value. Important Bird Area is a designation but without legal status.

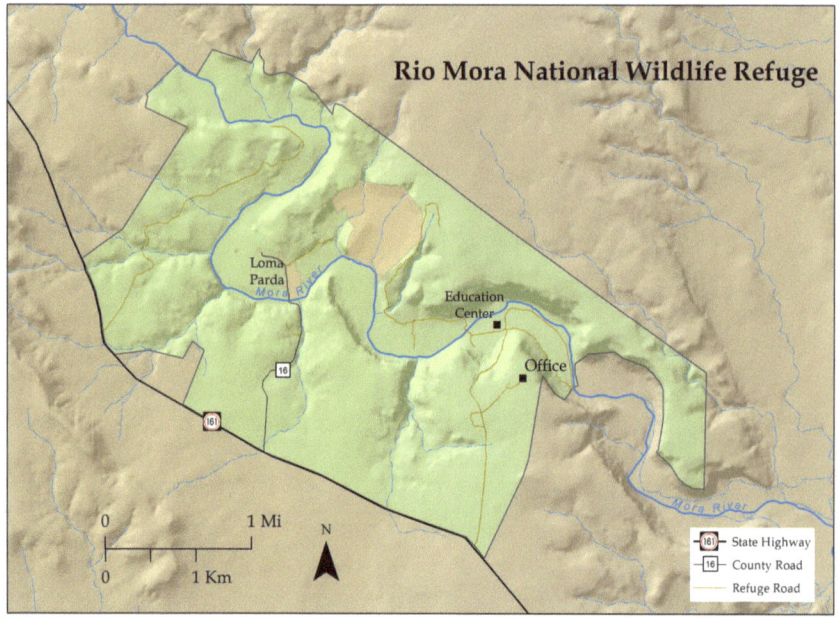

Figure 2.1. A map of the Rio Mora National Wildlife Refuge
Map courtesy of Joe Zebrowski

Habitats within the Mora River watershed provide important life cycle needs for a wide variety of neotropical migratory birds and many other riparian, grassland, woodland, aquatic, and wetland dependent species. The Migratory Bird Program in the USFWS Southwest Region has identified at least eighteen species from the Birds of Conservation Concern list that utilize the area during migration or for a winter stopover habitat.[14]

The refuge lies on the high plains east of the Sangre de Cristo Mountains. This is the southern end of the Rocky Mountain chain, a discontinuous series of ranges that extend from central New Mexico to northern Canada. They reflect red alpine-glow in the dawn. The elevation transition between the Great Plains and the Sangre de Cristo Mountains, which is the juxtaposition of two ecoregions, and the riparian habitats in this arid part of the West, all enrich the species diversity of the area. The location of the Rio Mora National Wildlife Refuge in the heart of this transition supports remarkable species diversity. We have documented 190 bird species, twenty-nine amphibian and reptile species, and fifty-one mammal species on the refuge (*see Appendix A*).

Figure 2.2. River valley at Rio Mora National Wildlife Refuge
Photo: Anabella Miller

About 140 million years ago, during the Cretaceous, eastern New Mexico was flooded by a shallow sea; this sea left thick deposits of shale and sandstone.[15] About eighty million years ago, the Laramide Orogeny began fault lifting Precambrian rocks upward to start the New Mexican part of the Rocky Mountain chain; the upward faulting continued into the Cenozoic era.[16] Along the east edge of the faulting, sedimentary layers bent upward to form the present-day hogbacks.[17]

Erosion from the mountains was heaviest during the Pleistocene epoch of the Cenozoic era because of continued uplift combined with ice-age precipitation. Dakota sandstone and Pierre shale still lie on the basin east of the mountains today. The dark gray Pierre shale was deposited as mud on the floor of the shallow sea; the Dakota sandstone, however, is a beach and shore deposit, and like beach sand it is porous and permeable.

Thus, the soil-covered sandstone serves as an aquifer throughout the east side of the Sangre de Cristo Mountains.[18] Various layers of soil cover this sedimentary base, with topsoil averaging about 4 inches (10 centimeters) thick.[19] Dominant soils are loam and clay.[20]

Up until about twenty-five million years ago, there were palms as far north as Montana. During the Miocene, from twenty-five million years ago until twelve million years ago, a drier climate forced the neotropical vegetation south and Nearctic vegetation north, and the grasslands emerged in between. During the Pleistocene, there were seventeen ice ages, roughly one every hundred thousand years. Vegetation moved south ahead of the ice, then moved north as the ice later melted—at least as a generalization of the time.[21]

Figure 2.3. Mammoth
Artwork: Brian Miller

At the end of the last ice age, the Clovis culture arrived in New Mexico and the rest of North America. The Clovis points were exquisitely beautiful as art and extremely deadly as weapons. Following human entry into North America, 70 percent of large animal species (averaging above 100 pounds (45 kilograms) fully grown) disappeared. Paul Martin developed the overkill hypothesis, and he was a brilliant man. In 1984, he assembled thirty-eight chapters into an edited book about the extinctions at the end of the Pleistocene. His chapter was called "Prehistoric Overkill: The Global Model."[22] In addition, two books were entirely devoted to why the ice age mammals declined: *Twilight of the Mammoths* by Paul Martin and *The Call of the Distant Mammoths* by Peter Ward.[23] A distant exception was the mammoth, which continued life on Wrangel Island in the Arctic Ocean until four thousand years ago. That is when humans arrived, coinciding with mammoth extinction on the island.[24] At every point around the world, when humans first arrived there were extinctions. Another example is the extinction of giant moas and marine mammals when humans arrived in New Zealand.[25] Another fascinating book about the ecological history of North America and its people is *The Eternal Frontier* by Tim Flannery.

Humans were new predators when first entering these places, so prey didn't have an antipredator response to them. As one person speculated, it was like putting a cat in a room full of canaries who had never seen a cat before. The tameness allowed humans to approach closely. And you can't blame humans for taking such prey. Humans had evolved as hunter-gatherers who had been subsidizing themselves by hunting for eons. The early entries into North America were simply trying to put groceries on the table. Over time, the humans of early North America equilibrated with nature. Importantly, the people of North America didn't domesticate large draft animals as did the peoples of Europe and Asia. According to Dan Flores, for that reason the people of North America didn't develop antipathy toward predators as did those in the Old World.[26] Respect for predators allowed ecological processes to continue, and bison were a constant source of food and hides. Predation, both human and wolves, didn't dent their numbers.

Modern bison and the other animals we associate with the grasslands appeared after the loss of the megafauna approximately twelve thousand years ago. The specialized Clovis and Folsom cultures were then replaced by a more generalized approach to life in the Southwest. About seven thousand years ago, the Altithermal period produced a two-thousand-year drought that depopulated the lowland areas.[27]

Ancient Pueblos lived in complex, multistoried buildings constructed of stone, adobe, and wood. The culture has existed for more than five thousand years. People often talk about places like Santa Fe and Jamestown being old settlements, but the Pueblo communities are far older. Both the Acoma Pueblo and Taos Pueblo have been consistently occupied for nearly a thousand years.

About twelve hundred years ago, Navajos and Apache arrived in the Southwest from the north. The Rio Mora NWR was once part of Jicarilla Apache lands and Pueblo lands. The Mora River offered a travel route for Pueblos coming down to the plains and plains tribes going up to the Pueblo villages to trade or raid. In addition to northern Pueblos and Apache, other tribes like the Comanche, Navajo, Ute, and Kiowa also passed through Rio Mora NWR, particularly after the arrival of the horse.

When the Spanish arrived, they subjugated tribes, but the Spanish were thrown out of Santa Fe by the Pueblo Rebellion of 1680.[28] Popay, who was from the Ohkay Owingeh Pueblo, united the Pueblo tribes, and they attacked the Spanish with 2,500 warriors. The Spanish returned in 1692 to retake Santa Fe.

The meadows where Las Vegas, New Mexico, is located today were given to Luis María Cabeza de Baca as a Spanish land grant in 1821, shortly before Mexico became independent of Spain. In fact, *vegas* means meadows in Spanish. The Baca Land Grant was used for grazing but was soon abandoned. The Las Vegas Land Grant from Mexico founded the town of Las Vegas in 1835, and the Rio Mora NWR was part of the Mora Land Grant from Mexico the same year. By this time the Santa Fe Trail was bringing US settlers into the area. In 1848, the Mexican-American War ended and the United States took control of New Mexico. Tribal wars continued with the US government.[29] The Apache Wars, Navajo Wars, and Comanche Wars were the most famous.

In the 1700s and 1800s, the grasslands of New Mexico were still very rich in quantity and quality. Grasslands were diverse, more productive, more resilient, and better able to absorb the impact of disturbances before European agriculture.[30] Soldiers and stockmen cut grama grasses for hay and claimed in their journals that it was of high quality. The diversity of grasses meant the prairie was better able to withstand drought and other disturbances.[31]

Today, the grasslands are very deteriorated and would be unrecognizable to a tribal elder from several hundred years ago. Most of this deterioration came after the Civil War. Overgrazing produced gully erosion, which lowered the water table and increased run-off after rains. Thus, the effectiveness of precipitation was reduced.[32] The elimination of beavers had reduced wetlands and access to moisture by grasses. Following 1870, the bison (*Bison bison*) were replaced with cattle. Bison had once numbered around thirty million and ranged across much of North America, but by the end of the nineteenth century, only about one thousand remained in the United States and six hundred more in northern Canada.[33] The slaughter of bison by hunters was encouraged by the US Army to subjugate warring tribes, to meet market demands driven by the sale of hides in the eastern United States, to provide bushmeat for forts and railroad workers, and to clear the land for homesteaders and domestic livestock.[34]

The 1870s were wet and grass was prolific. Although historical numbers vary, in 1870, there were 137,000 cattle in New Mexico; by 1880, the number of cattle increased to 500,000. During that same time, there were between two and five million sheep in northern New Mexico.[35] The wet 1870s turned into a decade of severe drought during the 1880s. Thus came the greatest livestock die-off in western history. Before dying, however, livestock denuded the grasses, and the subsequent erosion formed arroyos.[36] Indeed, some of the arroyos on Rio Mora NWR started during the 1880s. The overgrazing also started a juniper and piñon pine invasion of the grassland. Formerly confined to the rocky slopes, reduced grasses gave the trees a chance to sprout on the richer grassland soils. Unlike indigenous and Hispanic grazing practices, Northern European grazing practices suppressed fire. Bison, which break piñon and juniper on the grasslands, were replaced by cattle, which don't break trees. All of this contributed to the transition of western grasslands to savannas.

Early paintings of the Loma Parda ridge show grassy sides and a grayish color, hence the name—*parda* means gray in Spanish. The town was named after the ridge, and the ridge was a marker for people looking for the town of Loma Parda, which was in the valley formed by the Mora River. Loma Parda was originally settled as a farm town in the 1830s; but when Fort Union was built in 1851, the town became a place for off-duty soldiers to let off steam. Some called Loma Parda "Sodom on the Mora," and the local dance hall featured a series of small rooms in the back for the world's oldest profession. Legend has it that the church people of Las Vegas once came to Loma Parda to remove the prostitutes. In the process, they shaved the women's heads, but the women simply moved closer to the fort and set up shop in the caves of a canyon. That canyon, now part of Fort Union Ranch, is still called *Cañon de las Peloncitas* (Canyon of the Bald Women). Being closer to the fort may have simply improved business.

Today, the once-gray ridge is covered with piñon and juniper. Loss of vegetation allowed exotic plants to invade, including Russian thistle (*Salsola tragus*, commonly called tumbleweed), Siberian elm (*Ulmus Pamila*), Russian olive (*Elaeagnus angustifolia*), and cheatgrass (*Bromus tectorum*). Roads, fences, market hunting, predator control, and exotic species hastened the decline of wildlife. By 1900, grizzly bears (*Ursus arctos*), wolves (*Canis lupus*), and wild ungulates were so reduced in number that they bordered on extinction.[37] Game laws helped ungulates to recover, except for bison, but persecution of carnivores continues today, and that has removed the role of predators in top-down ecological processes. Top predators are keystone species, meaning they contribute far more to ecological processes than would be assumed from their numerical abundance. When carnivores are eliminated, the ecosystem degenerates. As in construction, remove the keystone and the arch collapses. This has been demonstrated in hundreds of ecological research projects.[38]

These are some of the legacy effects we still feel today, and they were a focus of restoration at Rio Mora NWR. We are fortunate that in this part of northern New Mexico many landowners, agencies, and NGOs are committed to changing those legacy wounds. I was fortunate to be a small help in that effort, and I took careful notes on what I saw, where I saw it, and what conditions supported it. I'll share those observations with you in the next chapters.

PART II

Field Notes on the Natural History of Rio Mora National Wildlife Refuge

CHAPTER 3

January and February

January 1 to February 28/29

Orion dominates the winter sky. During summer, he doesn't rise until it is nearly dawn. In fall, however, Orion becomes visible earlier and earlier as winter approaches. Seeing Orion early in the evening means cold weather is coming, and soon. January is a month of short days and long nights. The winter solstice has passed now, and the days are slowly getting longer. At this time of year, it is still hard to see much difference in daylight from one day to the next. Any heat stored in the ground over summer has been lost, so January and February can be bitterly cold on the high plains.

Birds

In early winter, mountain chickadees (*Poecile gambeli*) and Steller's jays (*Cyanocitta stelleri*) move down to the refuge from higher altitudes to spend the winter. So do a few goshawks (*Accipiter gentilis*), who cruise the piñon woodlands in winter. Black-capped chickadees (*Poecile atricapillus*), mountain chickadees, white-breasted and red-breasted nuthatches (*Sitta carolinensis* and *S. canadensis*), pine siskins (*Carduelis pinus*), piñon and scrub jays (*Gymnorhinus cyanocephalus* and *Aphelocoma californica*), and juncos (*Junco hyemalis*) are common at the refuge feeders in February. If you're lucky, you will see evening grosbeaks (*Coccothraustes vespertinus*) at the feeder as they move around the refuge. Their brilliant yellow and black color never ceases to amaze.

Birds can empty a feeder quickly in winter, but they don't eat all the seed they extract. They often cache seed in tree bark or in the ground for later feeding. Mountain chickadees weigh about a half ounce, but they survive harsh winters by collecting and storing tens of thousands of individual seeds in fall, which they find by memory during winter.[39] Mountain chickadees living in areas with harsher winters store more seeds than those living in areas with a gentler winter. In the far north, an individual can store as many as eighty thousand seeds for survival. They have one of the best memories of any vertebrate. The chickadees living in the far north have a larger hippocampus and larger neurons than the mountain chickadees living in less harsh conditions. In fact, females mating with males that have better memory have larger broods, so natural selection plays a key role.[40] No one knows how females assess a male's memory.

Around the first of February, bald eagles (*Haliaeetus leucocephalus*) that have wintered in New Mexico return to their nesting areas in the north. Bald eagle nests are rare in New Mexico. There are only about a half-dozen, but there is one high in a dead cottonwood tree on a neighboring ranch. In 2019, we first saw the nest when monitoring long-billed curlews (*Numenius americanus*). It was small for a bald eagle nest, which means the nest was fairly new. Old nests can be 8 or 10 feet (2 or 3 meters) across. During January, bald eagles add material to the nest for spring reproduction, then the female lays two white eggs in February.

Figure 3.1. Black-capped chickadee
Artwork: Mary Miller

Figure 3.2. White-breasted nuthatch

Photo: Dean Biggins

Figure 3.3. Bald eagle

Artwork: Brian Miller

Great-horned owls (*Bubo virginianus*) are common at the refuge. Actually, they seem to be at home almost anywhere and live in a wide variety of habitats. Their soft feathers insulate them against the cold nights. They are fierce predators that can take birds or mammals larger than themselves. Rabbits are tasty prey for them. The strong grip of their talons can sever the spine of their prey. They also take small mammals. I have seen rodent tracks that suddenly end with the mark of wingtips in the snow about 20 inches (51 centimeters) to each side of the last rodent track.

Mammals

While there is snow on the ground, the Botta's pocket gophers (*Thomomys bottae*) are still busy underground. Instead of just pushing up dirt into a mound like they do in the summer, the gophers burrow into the snow and pack those tunnels with the dirt they excavated when making their underground passageways. When the snow melts, you can see these cylindrical dirt tubes remaining on the surface of the ground. The castings are roughly the same width as the gopher's underground tunnels, about 3 inches (8 centimeters) in diameter, as shown in Figure 3.4.

Figure 3.4. Pocket gopher castings

Photo: Brian Miller

The underground burrow systems are complex with shallow side tunnels for food storage and latrines and deeper tunnels for nesting. Those nesting chambers are about 3 feet (0.9 meters) below ground. Gophers don't hibernate, and they eat roots year-round, which are supplemented by seeds and green vegetation. At the refuge, one gopher tunneled below a bird feeder at the office where it emerged early each morning to stuff its cheeks with as many fallen sunflower seeds as it could fit. With cheek pouches full of bounty, the gopher returned to its tunnel and backfilled the entrance as a precaution against snakes and weasels. Seeing this always brought a smile.

Figure 3.5. Botta's pocket gopher
Photo: Shantini Ramakrishnan

Voles live in subnivean (under the snow) tunnels, where they store food. Staying in those tunnels provides safety against the owl, although weasels (*Mustela* spp.) will enter the tunnels to hunt. Subnivean tunnels are also warmer than the air because a blanket of snow provides good insulation and moderates the temperature to around freezing, even when the temperature of the air is below zero. That's why Alaskan huskies on sled drives burrow into the snow to sleep. Humans can also use snow caves when winter camping.

Female black bears (*Ursus americanus*) give birth in January while hibernating in the den.[41] Hibernating black bears do not eat, drink, urinate, or defecate. They convert the nitrogen in urea into proteins that sustain their organs, including the brain. After extracting urea from urine, black bears recycle all water, and thus they avoid kidney failure. Because kidney failure is a cause of death for humans with severe burns, the medical profession is interested in research about how bears avoid kidney collapse during hibernation.

The two or three young are born weighing about 7 ounces (225 grams), the size of a rat. They are blind, hairless, and helpless, and they suckle their hibernating mother.[42] This is the ideal birth scenario for a female mammal, and it could make some humans jealous. Mom simply sleeps through the labor and the subsequent lactation. Mom and her young will emerge from hibernation in mid-April. According to Tom Beck, one of the leading black bear biologists in the nation, reproductive success depends on the fall berry and mast crop. If the crop is poor, pregnant females will reabsorb the blastocyst, and it will not implant into the uterus.[43]

Coyotes (*Canis latrans*), swift foxes (*Vulpes velox*), and gray foxes (*Urocyon cinereoargenteus*) also breed during January and February. Coyote gestation is about sixty-five days, gray fox is about sixty days, and swift fox gestation is about fifty days.[44]

Beavers (*Castor canadensis*) mate in January, February, and March and are monogamous.[45] They are semiaquatic, social animals and are active all year, although they typically spend more time in their dens during cold months. A group of beavers includes two adults, their yearlings, and young-of-the-year. That usually means about five or six beavers in each group. They are territorial, and the whole family defends their area from other beavers. A stretch of river defended by a beaver group can be a half a mile (1 kilometer) or more. Gestation is about 105 days, and when a new litter is born, the two-year-old beavers disperse to look for open territory.[46]

Beavers construct stick lodges sealed with mud or dens in a bank for housing.[47] The dens provide a hospitable environment during cold snaps and protection from predators. They make their dams out of sticks, rocks, mud, branches, and other material to seal the structure. Dams are usually built after spring run-off or in the fall to prepare for winter. The dams regulate water levels to enhance runways, create storage areas for branches that serve as winter food, and provide secure housing in lodges or bank dens. If you have ever tried to remove a beaver dam from an irrigation ditch by hand, you will admire the tight structure of the beavers' work. While their hind feet are wide and webbed, their front feet are dexterous with five digits capable of grasping.

Beavers can make quick work of a tree. While constant gnawing on trees wears down their teeth, beaver incisors grow constantly and are self-sharpening. Beavers typically cut trees up to 8 inches (20 centimeters) in diameter but can cut larger ones. The trees are used in the dam, and much of the tree-cutting happens in the fall. Beavers weigh about 50 pounds (23 kilograms), so they have the strength to drag large limbs. Sometimes they dig canals on the land to help get large diameter limbs to the water.

Figure 3.6. Beaver
Photo: Likka Keivula via Shutterstock

Most beaver groups on the refuge use bank dens. The steep banks of the channelized river are too deep for a dam to inundate a floodplain. With such steep banks, the river becomes disconnected from its floodplain, so there is no room for a side pond with a lodge in the middle. However, we have had a stick lodge on a low piece of land at the edge of the river. The beavers cut a runway to the lodge so that there was an underwater entrance. The bank dens have an entrance below the water level, and that entrance leads to a feeding chamber and dry sleeping chamber.[48]

Food stashes are usually a flat mat of branches on the floor of the river, sometimes under ice. The staple foods of beaver on the refuge are aquatic plants, willows on the floodplain, and the soft, inner layers of cottonwood trees (*Populus* spp.) that grow at the boundary of the floodplain and first terrace. Making a living on such a diet requires microflora (bacteria) to help digest lignin in the caecum. The stomach and intestinal bacteria help, but it's still not enough for complete digestion. Beavers recycle the undigested remains by eating their own feces. While the thought of this may make some people cringe, the technical jargon sanitizes this process by calling it *coprophagia*. The second chance at digesting woody material increases efficiency of nutrient uptake. It is actually a common practice. Coprophagia also happens in rabbits and pocket gophers.

Cottonwoods are like ice cream to beavers. We have pole-planted cottonwoods (and willows) to restore the riparian bosque on the refuge; the best time to plant is late winter and early spring. My first year at Wind River Ranch, I wanted to pole-plant some more cottonwoods, but it was late in the season to do that. On a Friday and Saturday, I simply put the poles in holes down to the water table and filled them with dirt thinking I could come back on Monday and fence the trees individually. By avoiding fencing at the time of planting, I thought I could get more trees into the ground. Over the weekend, beavers cut all the unfenced poles I had planted on Friday and Saturday and hauled them away. I hadn't quite fathomed the mind of a beaver yet, and a year later I attached a trail camera to a tree near a beaver dam in the river hoping to get some pictures of their work. The beavers, however, were concerned about privacy rights, and they cut the tree with the camera to put in their dam.

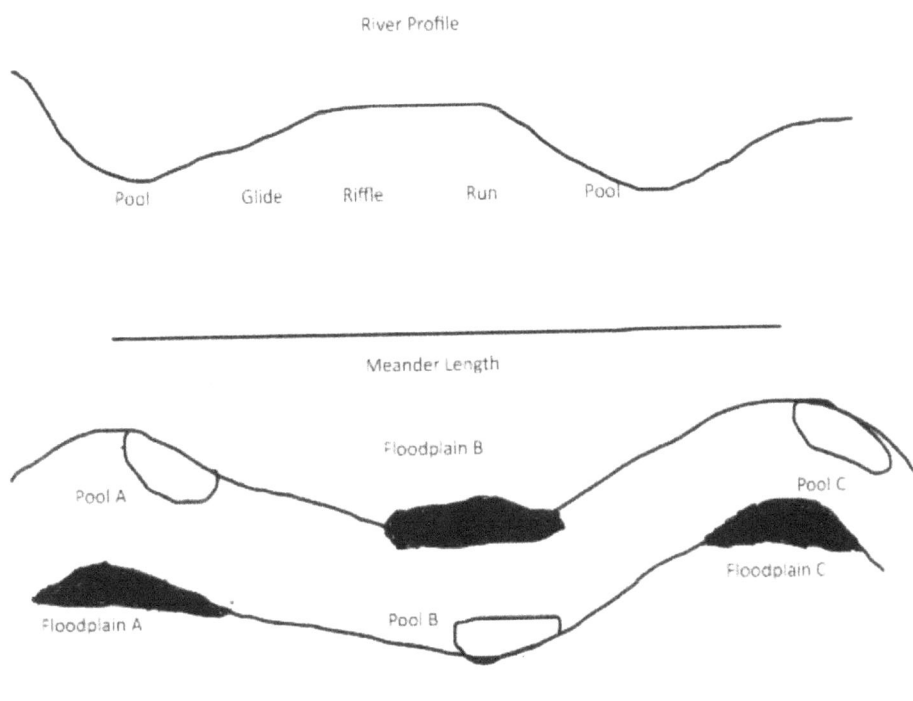

Figure 3.7. A meander schematic showing side-to-side meander pattern with pool and floodplain and the river floor undulation with pool, glide, riffle, and run sequence

Artwork: Brian Miller, modeled after Zeedyk and Clothier Let the Water Do the Work

Rivers have a side-to-side meander with a length between the top of one meander to the top of the next on that same side of the river, equal to about ten to fifteen times the width of the river at bank-full (*meander length* in Figure 3.7).[49] Bank-full width is the height of the river before it leaves its banks. The river floor has a sequence of pool, glide, riffle, and run (*river profile* in Figure 3.7). The pool is found at the outside bend of a meander. The inside bend forms a floodplain (*black* in Figure

3.7). The glide is at the exit of the pool. It is when you wade uphill in the river to move downstream. The riffle follows the glide and is usually in the middle of a straight flow from one meander to the next. It is the shallowest part of the river floor sequence and is the place where the meander moves from one side to the other. The run then forms the next pool. Over time, the river removes soil from the outside bend to form the pool and deposits it on the inside bend of the next meander to form a floodplain. For example, in Figure 3.7, the soil eroded from pool A is deposited on the inner bend of floodplain B. The erosion of soil from the outside of a pool means meanders of the river become more pronounced, but because that soil forms a developing floodplain at the inside of the next meander, the width of the river stays the same even though the meanders become more dramatic.[50] Rivers *want* to meander and need access to their floodplain. Understanding the sequence is a key to river restoration.

We see beavers on the refuge build dams at the riffle. This makes sense because the riffle is the shallowest part of the river floor. A dam at the riffle holds water to deepen the pool behind it. That makes bank dens and food storage easier. Any water that spills over the dam has accelerated force, increasing the scouring effect on the outside bend of the next pool. Thus, the activities of beavers promote the natural sequence of the river. Beavers maintain a connection between the river and its floodplain, and the floodplain recharges the water table. Willow growth on the floodplain slows the sheer force of high waters, reducing downstream flooding. Willows and cottonwoods stabilize the banks, and their roots form a net below the floor of the river, preventing channelization. The floodplain provides excellent wildlife habitat. A river that is disconnected from its floodplain is a river that has lost its ecological function.[51]

Beavers once numbered up to sixty million in North America.[52] When Europeans arrived, they trapped beavers to make felt hats to sell back in Europe. Beaver numbers plummeted under the pressure. Because beavers were such excellent workers, and the sound of running water led them to quickly fix breaks in a dam, the trappers would remove a bit of the dam and place a trap in the spot. The phrase "mad as a hatter" came about because they used mercury to make the beaver hats, and the hatters were exposed to mercury poisoning. Beavers were saved from extinction when the preferred men's hat changed from beaver felt to Chinese silk.

Today, the main force limiting beaver numbers is still human. While beavers manage watersheds and are thus keystone species—a species that contributes to ecosystem health in a manner that is greater than their numerical abundance—humans often only view them as clogging irrigation ditches, plugging culverts, and digging into canal banks. One of the reasons that keystone species, and other highly interactive species, are often persecuted is that the keystones push an ecosystem toward complexity and resilience. Unfortunately, for humans to profit economically, they need to simplify a system to control it.

Figure 3.8. Meander and riffle

Photo: Anabella Miller

CHAPTER 4

March

March 1 to March 15

Birds

In March, you will see mountain chickadees, Steller's jays, and goshawks return to higher altitudes. The first robins (*Turdus migratorius*), boat-tailed grackles (*Quiscalus mexicanus*), and red-winged blackbirds (*Agelaius phoeniceus*) arrive in early March, although boat-tailed grackles are in Las Vegas, New Mexico, during winter.

Mountain and western bluebirds (*Sialia currucoides* and *S. mexicana*) leave large groups and begin to pair in March. Bluebirds are cavity nesters, and they frequently use nest boxes. Western meadowlarks (*Sturnella neglecta*) are not yet singing, but they are starting to pair. Paired ravens (*Corvus corax*) dance on March winds in courtship, twisting, turning, and barrel-rolling like barnstormers in courtship flight. Take a moment to watch this acrobatic display.

Great-horned owls are nesting and sitting on eggs. The female will incubate for thirty-five days while the smaller male delivers rabbits, woodrats, and birds to her.[53] In this area, many nests are on protected ledges in the cliffs, although they also use stick nests in trees. Great-horned owls are one of the earliest nesters, and they often settle into old, existing nests of other raptors or ravens. The owls are formidable opponents, so other raptors do not try to move them out of a nest when they arrive from the south.

Golden eagles (*Aguila chrysaetos*) also nest in March, but mating occurs about a month earlier.[54] They often build their nests, called eyries, on steep cliffs that are out of reach for mammalian predators. They may also nest at the top of tall cottonwood trees. The eggs incubate for forty to forty-five days, and a pair may raise one or two young per season. They begin reproducing at five years of age and sometimes mate for life. Once highly persecuted, as many as twenty thousand golden eagles were killed in Texas and New Mexico between 1940 and 1960.[55]

Figure 4.1. Raven
Photo: Dean Biggins

The Evolution of Flight

Eggs are an avian adaptation for flight, as are hollow bones, and all species of birds reproduce by external eggs.[56] To fly, birds need to have a light body so that they can lift into the air without an excessive burst of energy. Laying eggs means that females can minimize the amount of time that they need to carry young inside their bodies, thus minimizing the amount of time that flying is made more difficult by that extra weight.

Lift is essential to flight because it gets the bird off the ground and into the air. Lift happens when air flows over a wing.[57] The curve of the wing forces the air to move farther over the top of the wing than across the bottom. Because flight requires that air flowing over the top and bottom of the wing reaches the back of the wing at the same time, the air flowing over the top of the wing moves faster than the air under the wing. Air moving faster over the top of the wing means air pressure above the wing is lower than below it, thus producing lift.

Humans applied the same principle to the wings of airplanes. The movie *Inherit the Wind* had a scene where Spencer Tracy mused on progress. He said, "Man learns to fly, but birds lose some of their majesty and the skies smell of gasoline." Both vertebrate and invertebrate fliers do so as part of an ecosystem. Humans fly at the cost of pollution, climate change, and transport of exotic species.

Archaeopteryx is the intermediate form linking reptiles and birds. It was flying 150 million years ago. Insects, however, evolved flight about four hundred million years ago. Bats (*Chiroptera*) began flying in the Eocene, about fifty million years ago.[58] Their wings are modified forelegs with elongated digits supporting a leathery membrane. Ancestral bats evolved from arboreal insectivores (today, insectivores are shrews). Thus, the first bats ate insects. Even today, about 70 percent of the bat species eat insects, including mosquitos. Bats are the only mammal that flies, and they are very good at it.

Flight is an example of convergent evolution, which occurs when two unrelated species develop similar traits or structures to occupy similar niches. Similar selective pressures produce similar traits in the unrelated species. These traits are called analogous traits.[59]

In contrast, divergent evolution occurs when species have a common ancestor, but over time become increasingly different. Traits in divergent evolution are called homologous traits because they are based in a common ancestor. As evolution forms new species from the common ancestor, an ancestral trait can be modified to fit new selective pressures.[60]

For example, the mammalian forearm evolving into a whale fin or a bat wing over time.

An Evolutionary Agreement between Jays and Pines

Piñon jays are highly social beings. They wander around in flocks and nest colonially. These flocks can include hundreds of birds. Piñon jays are named for their mutualistic association with piñon pine. Like the Clark's nutcracker (*Nucifraga columbiana*) in the high mountains and their mutualistic association with whitebark pine (*Pinus albicaulis*), limber pine (*P. flexilis*), and bristle cone pine (*P. longaeva*), the piñon jay has signed an evolutionary agreement with piñon pines.[61] The pine nut feeds the jay, and the jay stores the pine seeds in caches far from the mother tree, acting as an agent of dispersal. In return for food, the piñon jay thus helps the piñon pine reproduce. The strong bill of the piñon jay can hammer open the cone and extract the nut. After a group visits a tree, the broken pieces of the cone are scattered around the ground near the tree. Scrub jays have a weaker bill and are not very good at exploiting the resinous cones of the piñon pine. The piñon jays have excellent spatial memory and can find the caches throughout the winter, even under snow. Because of the caching, piñon jays nest in winter. Nestlings eat insects in spring.[62]

Figure 4.2. Piñon jay
Photo: Rich Reading

Pines that live in fertile soils disperse their seeds by wind. When drying, the interior fibers of these cones shrink, which exposes the seeds to the air. Then, the breeze carries them away. Those seeds are likely to land where germination can take place.

Pine species, like the Piñon pine in New Mexico, that grow on poor soil are called stone pines. They pack nutrition in the form of fatty acids around the seed to aid germination.[63] A pound (0.5 kilograms) of piñon nuts contains nearly three thousand calories, which is more than a pound (0.5 kilograms) of chocolate.[64] The extra weight of the seed means wind is of little use for dispersal, and that is where the jays come into the picture. If the seed fell only below the mother tree, the chances of germination would be slimmer. Usually, piñon jays carry nuts 1 to 3 miles (2 to 5 kilometers) from the mother tree, but they can carry them up to 7 miles (11 kilometers) away. Some seeds are left after the cache is opened, and those seeds can germinate. A single piñon jay may cache twenty thousand seeds in a season, and ecologist David Ligon estimated that a flock of 250 birds can cache thirty thousand seeds a day and 4.5 million seeds in a season.[65] Much of what we know about piñon jays comes from David Ligon and his students at the University of New Mexico.

Figure 4.3. Red-winged blackbird
Photo: Anabella Miller

Figure 4.4. Mountain bluebird
Photo: Anabella Miller

Mammals

You now can see mature elk (*Cervus canadensis*) dropping their antlers, but younger males will still have theirs, particularly spikes who shed later in the spring. Deer mice and white-footed mice of the *Peromyscus* genus begin breeding in March and can continue to breed until October.[66] Gestation is about four weeks, and a female can breed at eight weeks of age, so a mouse born early in the year can contribute to reproduction of the species later in the year. They eat almost anything and can live almost anywhere, so they are very common. They rarely live long because they are prey to so many species.

Deer mice (*Peromyscus maniculatus*) are one of the main carriers of hantavirus.[67] Rodents shed the virus in urine, feces, and saliva. When that material is disturbed, tiny droplets of the virus rise into the air; humans can get the virus from inhaling those droplets. If a human gets the virus, it becomes a serious lung disease called hantavirus pulmonary syndrome, which can be fatal. About one out of three people who are infected don't make it. If you are cleaning in a confined area that has evidence of rodents, spray the area with a solution of 90 percent water and 10 percent bleach. Try to keep your home clean of rodents.

Plants

Chances are that the first green growth of the year you see in the valley at the refuge is poison hemlock (*Conium maculatum*), but storksbill (*Erodium cicutarium*) leaves also green early. A native of Eurasia, storksbill was introduced into North America in the eighteenth century and is now naturalized in the Southwest. Poison hemlock, a member of the carrot family, is also a non-native from Eurasia that is now widespread in North and South America.

Poison hemlock is the most toxic plant we have in the area.[68] Supposedly, it was in the elixir that killed Socrates in 399 BC. You can be infected simply by brushing the plant with a sweaty arm. Even very small amounts can kill you in a couple of hours. Toxicity varies depending on the part of the plant consumed and the time of year. It's most toxic in the spring, and the root is more toxic than the leaves. Death comes by respiratory failure due to paralysis of respiratory muscles. You can survive if you are intubated until the poison is flushed from your system and your lungs can begin respiration again. If you are trying to remove poison hemlock, wear gloves and eye protection. Do not burn it, as the fumes are also poisonous. Do not compost it, as the poison can last up to three years after plant death. Put it in a plastic bag.

In early March in Rio Mora, one-seed juniper (*Juniperus monosperma*) and Rocky Mountain juniper (*Juniperus scopulorum*) pollen is in the air. Many people (including my wife) have an allergic reaction to those tree species. Siberian elm is also releasing pollen, but that tree is exotic to the area. It is common in Las Vegas and Santa Fe.

March 16 to March 31

Spring Equinox

At last spring has arrived. The spring equinox marks the first day of spring in the northern hemisphere. This is the day that the sun crosses the equator, balancing the day with twelve hours of daylight and twelve hours of night. From now until the summer equinox, the days are longer than the nights. The first thunder can come any time now.

Seasonal variation is caused by the spherical shape of the Earth and the angle of the Earth as it rotates on its axis while orbiting the sun. The axis angle of rotation is not perpendicular but is tilted about 23.5 degrees away from perpendicular. This tilt is unchanged throughout the orbit. So, the solar energy hitting the northern and southern hemispheres changes as the seasons change. The sun is directly over the equator at the spring and fall equinoxes (about March 21 and September 21), but at the summer solstice in the northern hemisphere, the sun is directly overhead at 23.5 degrees north of the equator. Thus, the northern hemisphere gets longer sunlight, giving us longer and warmer days. At the northern winter solstice, the sun is directly overhead at 23.5 degrees south of the equator, and our northern days are short and cold. Temperatures between 23.5 degrees south and 23.5 degrees north don't change much throughout the year, although precipitation patterns do. Solar energy heating the surface and atmosphere drives air circulation and influences precipitation patterns.

Birds

The first waterfowl pass through, but they are still in low numbers. Robins, red-winged blackbirds, and boat-tailed grackles increase their numbers, and the first killdeers (*Charadrius vociferus*) of the year announce their arrival with their eponymous call. The songfest begins to waft throughout the refuge. House finches (*Carpodacus mexicanus*), robins, bluebirds, and western meadowlarks are all singing.

It is not hard to distinguish the songs. The house finch warble lasts for three or four seconds with the last note higher than the rest. Robins have a slow, evenly spaced warble that says *cheerily, cheery up, cheery-oh*. The meadowlark song is a rich and loud *Chilé-con-tor-TEE-ahs*. Townsend's solitaires (*Myadestes townsendi*) have switched from their winter melody to the high-pitched single *cheep* of warmer weather. Our ant-eating northern flickers (*Colaptes auratus*) are drumming and giving their rapid *week-week-week* call. They also make the single note call that sounds a little like a female elk calling her young. To advertise spring, flickers drum on surfaces that resonate and accent the noise. Drainpipes on houses will serve that purpose well. By mid-March, we can also see a few Say's phoebes (*Sayornis saya*).

Figure 4.5. American wigeon
Photo: Bill Zeedyk

Figure 4.6. Mallards
Photo: Anabella Miller

Figure 4.7. Canada geese
Photo: Dean Biggins

Green-winged teal (*Anas crecca*), blue-winged teal (*Anas discors*) mallards (*Anas platyrhynchos*), American wigeons (*Anas americana*), redheads (*Aythya americana*), Canada geese (*Branta canadensis*), ring-necked ducks (*Aythya collaris*), northern shovellers (*Anas clypeata*), gadwalls (*Anas strepera*), ruddy ducks (*Oxyura jamaicensis*), common mergansers (*Mergus merganser*), buffleheads (*Bucephala albeola*), and snow geese (*Chen caerulescens*) head north in successive waves, each wave with large numbers. Many of our waterfowl winter at the Bosque del Apache National Wildlife Refuge, which is south of Albuquerque—a must visit to enjoy waterfowl during winter. Spring migration is a critical time. Waterfowl must arrive at their destination with enough energy to breed. All migrating birds headed north need stopover places to refuel. During the fall migration, they only need to return to the wintering area, which requires less energy than reproduction, although many small migrating birds heading south need to replenish energy during frequent stops. That is the value of national wildlife refuges with ponds as well as flooded fields and seasonal floodplains carrying spilled grain and plant detritus. Waterfowl can stop and feed to build energy as they head north or south.

While ducks and geese move north, the ones that will stay here to nest are now in pairs. We always have several wood duck (*Aix sponsa*) nests, even though Rio Mora is out of their purported range. If you have not seen a wood duck, the male is one of the gaudiest and most beautiful of all the waterfowl. Mallards, American wigeons, northern shovelers, blue-winged teal, common mergansers, ruddy ducks, and Canada geese can and do nest here.

Sandhill cranes (*Grus canadensis*) fly over the refuge. Some stop to spend the night, but they nest farther north. When sandhill cranes get to their nesting area, the mated pair share a call. One gives a loud honk, and the second quickly adds the trill. Because the pair do this at a distance from each other, it can confuse a predator trying to locate the nest. The body is a graying tan with a bright red cap on the head.

Figure 4.8. Sandhill crane
Artwork: Mary Miller

Mountain plovers (*Charadrius montanus*) are also in the area. Some will nest in northeastern New Mexico, but populations have declined by 80 percent in recent times, largely because of habitat loss due to agriculture. Lost winter areas also hurt. One historical document named a winter area for mountain plovers as west of Los Angeles, California. For nesting, the female mates with a male, lays the eggs, then lets the male brood those eggs. She mates again, lays eggs in a second nest, and she broods those eggs herself.[69] Great blue herons (*Ardea herodias*) begin to nest around this time as well. They reuse rookeries, or colonial sites, where there are multiple stick nests in the same tree.

Figure 4.9. Great blue heron
Photo: Dean Biggins

The prairie falcons (*Falco mexicanus*) and peregrine falcons (*Falco peregrinus*) arrive at Rio Mora NWR in mid- to late March. The peregrine falcon nest is a scrape on a cliff ledge, usually a steep cliff near a river.

The prairie falcon scrape is also on a cliff, but it can be farther from water than the peregrine scrape. The peak for nesting is April and early May. Females of both species will sit on eggs for about thirty days while the male brings her food. The hatchlings may fledge in another thirty or forty days, and they can be independent at two months of age.[70] Figure 4.10 shows a peregrine falcon in flight. The prairie falcon has a similar outline while in flight. At a distance, however, you can see black armpits on the prairie falcon. The peregrine falcon has a thicker and blacker mustache, and the peregrine back is darker than the brown prairie falcon back.

Figure 4.10. Peregrine falcon
Photo: Harry Collins via Adobe Stockphoto

They may reuse the same scrape in following years, or they may use a different scrape in the same general section of cliff, as the ones in Rio Mora do. The courtship display of peregrines and prairie falcons is something to behold. The male soars high into the sky, so high that he almost disappears. Then he dives toward the ground at speeds nearing 200 miles per hour (mph). At the last minute he pulls out of the dive, turns upward, and returns to the heights for another dive. The female perches to watch the display. Because the male will feed her while she incubates the eggs, she wants to make sure she courts with a good athlete and reliable provider. If she is satisfied, they copulate on a prominent perch, often a rocky outcrop. She angrily reveals the nest if you come too close.

The Dangers of DDT

Peregrine falcons and bald eagles were decimated by dichlorodiphenyltrichloroethane (DDT) use in the middle of the twentieth century. DDT caused egg thinning because the pesticide interferes with calcium deposition in the egg shell.[71] The act of brooding crushed the shell. Peregrine falcons were affected, while prairie falcons weren't because peregrine falcons eat birds who eat insects. The pesticide moved up the food chain for them. Prairie falcons, however, eat mammals, and many of their prey eat seeds or vegetation. Prairie falcons are equally susceptible to egg-thinning by DDT, but by eating mammals they simply weren't exposed.

DDT was first developed in 1874 by Austrian chemist Othmar Zeidler, but it wasn't until the 1940s that it entered agriculture as an insecticide. Its insecticidal power was discovered in 1939 by Swiss chemist Paul Muller, and he was awarded a Nobel Prize in medicine in 1948 for this destructive discovery. The government and industry promoted DDT, and it was made available to the public. From the beginning, however, there were environmental concerns. In 1945, Rachel Carson began studying the effects of DDT on wildlife. In 1962, she published *Silent Spring*, a book about her findings on the negative effects of DDT. While her bravery sparked an environmental movement and led to the United States banning DDT in 1972, she paid a price. The chemical and agricultural industries vilified her to the public. A worldwide ban on DDT was formalized in 2004 via an environmental treaty by the Stockholm Convention. During the 1950s, the United States put DDT in barrels and dumped them off the California coast.[72] They assumed it would degrade into something harmless. They were wrong. Recently, people investigated the barrels and discovered the DDT is still potent.

Mammals

Bees that nest below the ground, like the family Halictidae, commonly called sweat bees, begin to buzz in their tunnels. Sometimes you can see the tracks of a skunk who heard the buzzing and scraped down into the soil to get a meal. The skunk digs with its long front claws and roots with its snout to get these insects. There are three species of skunk on the refuge: striped skunk (*Mephitis mephitis*), western spotted skunk (*Spilogale gracialis*), and the hog-nosed skunk (*Conepatus leuconotus*). The hog-nosed skunk is common in Mexico, but we are at the northern extent of its range.

Figure 4.11. Western spotted skunk
Artwork: Mary Miller

All skunks have warning coloration to dissuade predators, although skunks are shy and not aggressive. The black-and-white markings warn an intruder that they can shoot a sticky, putrid spray that will linger on the recipient. They will display discontent with warning signs, such as foot-stomping, pawing, and hissing, before they release their anal scent glands. The acrobatic spotted skunk will plant its front feet on the ground and do a handstand as a final warning. They shoot the spray from two nipples on either side of the anus. The nipples can be rotated individually to aim and are accurate to at least 10 feet (3 meters).[73] They aim at least one nipple toward the eyes of their target, and the effect is much like tear gas.

In late March, male rock squirrels (*Spermophilus variegatus*) and male Gunnison's prairie dogs (*Cynomys gunnisoni*) begin to emerge from hibernation. Animals come out of hibernation when brown fat metabolizes to increase body temperature and heart rate. Brown fat produces more heat than white fat because brown fat contains more mitochondria.[74] Females will emerge in a few more weeks, but the young of last year stay below ground until late April.

Figure 4.12. Rock squirrel
Artwork: Brian Miller

Plants

Juniper pollen is still high. We are headed toward warmer spring temperatures, and cottonwood buds are growing. We can, however, still have a late blast of winter. This will not injure the cottonwood buds. If there is a cold snap, they simply arrest development until weather is more suitable. By mid-March, locoweed leaves (*Oxytropis* spp.), Easter Daisy leaves (*Townsendia excapa*), and pasque flower (*Pulsatilla patens*) leaves turn green. Storksbills are one of the first spring flowers. They are an exotic from Europe, but they do quite well in New Mexico.[75]

Insects

Like the Halictidae bees, harvester ants also winter below ground. As a community they huddle around their queen. During winter, they close the openings to the surface and become inactive. As the weather warms, they regain activity below the surface. If you break the mound near the first day of spring, you will find them moving, although not yet above the surface. In summer, the mound looks harmless until you poke it. Ants are an important source for moving seeds, thus helping plant dispersal.[76]

On the first days of spring, you can spot a mourning cloak butterfly (*Nymphalis antiopa*) or a Horace's duskywing skipper (*Erynnis horatius*). Mourning cloaks spend the winter as adults under loose tree bark or in a tree cavity. This is unlike most butterflies who overwinter as caterpillars or pupae. Mourning cloaks can emerge on an early warm day, but if cold returns they go back to their overwinter shelter. When they finally emerge from their hibernacula, they breed and lay eggs on their host plant—often willows or cottonwoods. The metamorphosis process then begins again.[77] They are territorial and defend their space. Most of the wing is a dark brown with a yellow edge. There are blue spots on the brown at the border with the yellow. Duskywing skippers overwinter as a fully grown caterpillar. On warm days, syrphid flies (*Syrphidae*), commonly called flower flies, dart about feeding on nectar.

Figure 4.13. Mourning cloak
Artwork: Brian Miller

CHAPTER 5

April

April 1 to April 10

Birds

Birds seem to arrive at roughly the same time each spring. The turkey vultures (*Cathartes aura*) that roost here arrive reliably around the first of April. They breed in April and May and lay eggs in a scrape on the ground or on a ledge in areas where humans are scarce. They incubate for a month, then the young are in the nest for another two months. They are graceful fliers who glide on air currents for long periods between wing flaps. From a distance, you can see a slight V shape of the wings as they tip from side to side.

Nearly all juncos have switched from their winter chipping to their spring call, a high-pitched *trill*. Chipping sparrows (*Spizella passarina*), which won't arrive until nearer to the end of April, have a similar trill, although it is flatter in tone. You can hear the Say's phoebes singing a quick *PEE-you* sound. Look for them perching on twigs as they hawk for insects. Some migrators, like willow flycatchers (*Empidonax trailii*) and Lincoln's sparrows (*Melospiza lincolnii*) stop at the refuge in early April to rest on their way north.

Figure 5.1. Say's phoebe
Photo: Anabella Miller

Figure 5.2. Downey woodpecker
Photo: Anabella Miller

Downey woodpeckers (*Dryobates pubescens*), the smallest of the woodpeckers, are commonly seen year-round residents. They resemble the slightly larger hairy woodpecker (*Leuconotopicus villosus*), although the hairy woodpecker has a longer bill. Both use stiff tail feathers to prop against trunks and branches of trees. Both nest in cavities that they begin to make about two weeks before laying eggs. Males and females of both species bring insects to the young. They eat beetles living in bark and wood, ants, caterpillars, and many species that humans consider pests. You also see the first double-crested cormorants (*Phalacrocorax auratus*) flying over, although some nest in northeast New Mexico.

Plants

You can now see white Easter daisies and purple pasque flowers blooming. They keep a very tight schedule for appearing near April 1, but other flowers (and insects) can be thrown off by weather or precipitation changes from one spring to the next. When you see the white Easter daisies and purple pasque flowers, you know you have made it to spring. *Pasque* is French for Easter, which comes close to the time of the blooms.

Notice how the buds of the cottonwood and boxelder (*Acer negundo*) are swelling. You'll see both Fremont cottonwood (*Populus fremonti*) and narrowleaf cottonwood (*P. angustifolia*) along the Mora River on the refuge. Storksbills begin blooming now with pinkish-purplish flowers.

Cool season grasses are greening. The green color comes from chlorophyll that converts energy from sunlight into carbohydrates by the process of photosynthesis. During photosynthesis, plants absorb carbon dioxide and water to make carbohydrates, called carbon fixing. In the process, the plants give off oxygen as a byproduct. Herbivores eat carbohydrates in plants, and carnivores eat the herbivores. As carbon moves up the food chain, animals breathe in oxygen to make energy for cells. We call this process respiration. While producing energy during respiration, animals exhale carbon dioxide as a byproduct. So, photosynthesis and cellular respiration are inextricably linked. The products of one are the reactants of the other, and this perpetual cycle sustains life on Earth.

A fixed amount of energy comes to Earth in the form of sunlight, but humans have been able to expand our energy use beyond the limits of daily sunlight by harvesting past sunlight in the form of fossil fuels. Fossil fuels, however, are not renewable in anything close to human time scales. Estimates are that we will expire fossil fuels and natural gas in around fifty-five years, plus or minus, while coal could last for 110 years.

Insects

On a warm spring day, there are duskywing skippers, mourning cloak butterflies, and early grasshoppers, like the speckle-winged grasshopper (*Arphia conspersa*), the crowned grasshopper (*Trachyrhachys coronata*), and the red-shanked grasshopper (*Xanthippus corallipes*), and plenty of tiny lygaeisids (Lygaeidae, a family of the order Hemiptera), which seem to be everywhere, but particularly on the greening tansy mustard (*Descurainia pinnata*) plants and the willows (*Salix* spp.).

The early grasshoppers emerge, mate, and then lay eggs. The eggs hatch in the summer during the rains. The young mature through the instar stages, then overwinter as instars that are nearly adult. Thus, they are one of the first grasshoppers of spring. There is one generation a year. Those early grasshoppers are fuel for migrating birds as well as the long-billed curlews that nest here. In riparian areas, you can see green-striped grasshoppers (*Chortophaga viridifasciata*) that do not inhabit the expansive, upland prairies.

Mayflies (family: Baetidae) fly in the first week of April, and painted lady butterflies (*Vanessa cardui*) arrive soon after mayflies appear. You will see a painted lady here and there, but in 2008 and 2020, there were very large numbers flying through, perhaps thousands. People once thought that painted lady butterflies who made it from Mexico to Alberta, Canada, died there in the fall. Recent research in Europe, however, showed that painted lady butterflies arriving in Finland did not die in the fall. Instead, they rose high in the atmosphere and rode the air currents back to Africa. The same may happen in North America. You can see plume moths (*Hellinsia homodactyla*) as appear in the April evenings. The first harvester ants—called pogos for their genus *Pogonomyrmex*—come up to the surface. Ants on the surface are prime food for northern flickers.

Figure 5.3. Painted lady butterfly
Photo: Dean Biggins

April 11 to April 20

Birds

Geese and ducks begin their nesting activity. Canada geese prefer nesting on small islands in the river. The female incubates the eggs for twenty-eight days. You can hear turkeys (*Meleagris gallopavo*) gobbling. Robins begin making nests. Golden eagles, bald eagles, and great-horned owls now have hatched chicks, or soon will. Some migrating birds, like tree swallows (*Tachycineta bicolor*), begin to pass through on their way north, snatching some late mayflies from the air.

Horned larks (*Eremophila alpestris*) are now in pairs, and you should hear song sparrows (*Melospiza melodia*), vespers sparrows (*Pooecetes gramineus*), and spotted towhees (*Pipilo maculatus*) singing. The vesper sparrow song sounds like *here, here, there, everybody-down-the-hill*. For some reason, this song is one of my favorites. The song sparrow says *hip, hip, hip, hooray boys, the-spring-is-here-again*. The spotted towhee song is *here, here, here, PLEEEEASE*. They also make a noise that sounds like *churrr*. The first Lewis's woodpeckers (*Melanerpes lewis*), ferruginous hawks (*Buteo regalis*), and Swainson's hawks (*Buteo swainsonii*) appear. Rough-winged swallows (*Stelgidopteryx serripennis*) fly over the refuge heading north, although some do nest here. Long-billed curlews return around the tenth of April.

Figure 5.4. Horned lark
Artwork: Anabella Miller

Figure 5.5. Great-horned owl
Artwork: Mary Miller

Figure 5.6. Spotted towhee
Photo: Anabella Miller

You can now see loggerhead shrikes (*Lanius ludovicianus*). They are listed as wintering in New Mexico, but we don't see them during the cold months. Red-tailed hawks and falcons are busy incubating their eggs. The red-tailed hawk scream is famous and often used in movies, even if the movie location is nowhere near the red-tailed hawk range.

Mammals

Chipmunks are active now. They don't sleep the entire winter. Instead, they awaken every few days to eat from their winter cache because they don't have the high levels of fat that hibernating rodents have. On the refuge, we have mostly Colorado chipmunks *(Tamias quadrivittatus)*, but there are a few least chipmunks *(Tamias minimus)*. When running, least chipmunks hold their tail straight up in the air, whereas the Colorado chipmunk leaves its tail trailing behind the body. Thirteen-lined ground squirrels (*Spermophilus tridecemlineatus*) emerge from hibernation. Adult males are only active about three months of the year. During hibernation, the heart rate of thirteen-lined ground squirrels drops to about four beats a minute. While active, the heart rate is two hundred beats per minute. Up to half the diet of these ground squirrels can be animal matter, such as roadkill and insects, with the other half being plant matter.

Insects and Plants

The primary blooming flowers that you see in mid-April are storksbill and dandelion. Painted lady butterflies prefer the dandelion. You can see them flitting from one yellow flower to the next. Horace's duskywing skippers and a few variegated fritillary butterflies (*Euptoieta claudia*) appear on the grassland. Green-striped grasshoppers, speckled winged grasshoppers, crowned grasshoppers, and red-shanked grasshoppers are still active as well. The tiny lygaeisids begin to drop in number, and the tansy mustard is about to seed. Invasive Siberian elm starts to look green, and they are easy for you to spot because of the color of the seeds that they eject.

Reptiles and Amphibians

Cold-blooded reptiles don't synchronize with other events, as they react more to temperature. Bull snakes (*Pituophis catenifer sayi*), prairie rattlesnakes (*Crotalis viridis*), and garter snakes (*Thamnophis* spp.) were up and about at this time in 2018, but that was a month earlier than they emerged in 2019. When it is warm enough, you will see them. If you do see a garter snake, don't worry. They are harmless. Their main defense is a foul-smelling secretion from musk glands. Their saliva is toxic to rodents, but harmless to most people. Plus, they rarely bite.

Bull snakes can produce an effective imitation of rattlesnakes, even shaking a tail, but they have no venom.[78] Bull snakes may bite, however, if handled brusquely. While imitating a rattlesnake may be an effective ruse to fool the bull snake's predators, it is ineffective against humans who fear rattlesnakes and can't tell one from the other. While the markings on the body are similar between the western rattlesnake and the bull snake, the heads are different (Figure 5.7). The rattlesnake has a broad, triangular head, which accommodates its large fangs, poison glands, and sensory organs. Bull snakes, however, have a gracile head, which can help you tell the two species apart. If you can't positively identify a snake, don't touch it.

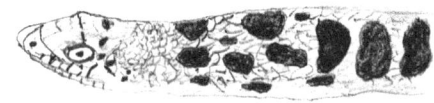

Figure 5.7. Heads of a rattlesnake (*top*) and a bull snake (*bottom*) for comparison

Artwork: Brian Miller

There is a folk tale that says bull snakes prey on rattlesnakes and can reduce rattlesnake numbers. While diet studies have shown a few rattlesnakes in the bull snake diet, it is only about 0.01 percent, and that is far too low to influence rattlesnake numbers.

Rattlesnakes have an interesting adaptation in cold climates. The female can mate, then store sperm over the winter. That sperm is still viable next spring.[79] There is also a saying that a dead rattlesnake can still bite. That is true. After death, the rattlesnake nerve endings can still respond for a period of time. The brain may be dead, but the muscles, nerves, and quick bite response are still functional for several hours. You can even be envenomated by a decapitated rattlesnake head. Thus, if you find a dead rattlesnake, don't pick it up.

Figure 5.8. Leopard frog
Photo: Carmen Briones

Male leopard frogs (*Lithobates pipiens*) start to sing to attract females, who are quiet. Males make a sound resembling a low guttural and clicking growl.[80] Some say it sounds like rubbing your finger on a balloon. The song doesn't carry very far. April and May are the prime months for breeding. Females can lay two thousand to four thousand eggs. Leopard frogs are both predators of insects and prey for other predators like aquatic snakes and medium-sized mammals. Their defense is the camouflage of their spots when in vegetation. They play a role in nutrient cycling. Leopard frogs have been declining due to chytrid fungus (*Batrachochytrium* spp.), a skin disease affecting amphibians. Chytrid could be a contributing factor to the global decline in many amphibian populations.[81] It has affected about 30 percent of the amphibian species of the world and is a serious problem in the area in and around the refuge.

Competition with bullfrogs (*Lithobates catesbeianus*) has reduced leopard frog numbers in the west.[82] In the eastern United States, bullfrogs and leopard frogs coexist because there are more native predators that prey on bullfrogs. That predation erects ecological boundaries that limit bullfrogs and allows leopard frogs to thrive. In the western United States, those eastern predators are mostly absent, so bullfrog numbers have proliferated. There is also more water in the midwestern and eastern United States and a greater diversity of aquatic habitats. Bullfrogs are native to the Mississippi watershed, but they inhabited lower elevation wet habitats of the watershed. There are no records of bullfrogs in Colorado before 1800.[83] Thus, when bullfrogs were introduced to the high plains of New Mexico, the fewer bodies of water put bullfrogs in greater contact with other frogs.

Bullfrogs are the biggest frogs in North America. If you grab one, they can be more than a foot (0.3 meters) long from nose to their hanging legs. They can pretty much eat anything they can fit into their mouth, including large prey items, because of their strong bite. They are ambush predators. The quickness of their tongue flick comes because when a bullfrog closes its mouth, there is tension on the elastic parts of the tongue.[84] Opening the mouth releases the tension, and the tongue shoots out like a slingshot. So, tongue movement is more than just muscle, and it moves faster than the escape response of the prey. This elastic response means that the bullfrog can capture prey in cold water as well as warm. An experiment at the refuge to reduce bullfrog numbers has helped leopard frog survival. Reintroducing river otters (*Lutra canadensis*) to their former range could also help in areas where bullfrogs were introduced.

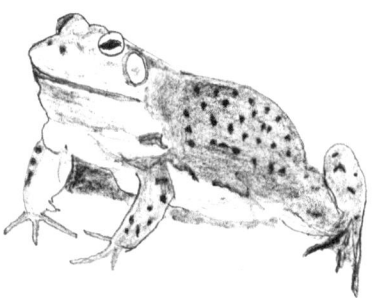

Figure 5.9. American bullfrog
Artwork: Brian Miller

Bullfrogs were introduced to Rio Mora NWR by one of the former owners of the ranch who was from Texas. He liked to eat them, and they are tasty. April and May are the breeding season, and the period of sexual activity is prolonged. Male bullfrogs can congregate in groups at a breeding pond. Their group singing is analogous to lekking birds where the female chooses a mate. Females can lay twenty thousand eggs in a mass, so it is easy to see how introduced bullfrogs can negatively affect native species.

Bullfrog tadpoles can take two years to change into frogs, and older tadpoles can get quite big. At Rio Mora NWR, they can grow to 4 or 5 (10 or 13 centimeters) inches in length. During winter, we see them swimming under the ice. When bullfrog tadpoles change into frogs, the process can go quickly once the front legs appear on the tadpole. The thyroid hormone plays a critical role in metamorphosis.

Western chorus frogs (*Pseudacris triseriata*) will sing when the rains make ephemeral ponds.[85] Using permanent bodies of water exposes them to higher rates of predation. They can sing quite early at high altitudes and can withstand temperatures of 23 degrees Fahrenheit (-5 degrees Celsius). They can survive the cold because they produce glycolipids, which prevent cell rupture when frozen.[86] When temperatures warm, the blood thaws and begins to flow as they enter an active state. Like the leopard frog, April and May are the prime breeding months for western chorus frogs, but during one drought year we didn't hear the mating chorus until the end of July.

April 21 to April 30

Birds

A Lewis's woodpecker was still hanging around headquarters during late April in 2019. A single bird during nesting season often means the other member of the pair is sitting on eggs. We saw multiple Lewis's woodpeckers nesting from 2005 to 2008, and there were multiple adults around each nest hole. It appeared to be cooperative breeding.[87] When nest holes are limited, the young from the previous year stay to help the breeding pair at this year's nest. The young from last year forego breeding to maintain contact with the cavity because it can be difficult to find suitable nest sites. It is more advantageous to help the breeding pair, and hope to inherit the nest hole, than it is to fly off searching for a new cavity. This cooperative breeding strategy has been documented in less than 10 percent of bird species, however, very few bird species are obligatory cooperative breeders—meaning there would be no reproductive success without it.

When feeding chicks, Lewis's woodpeckers are more like flycatchers, in that they hawk insects in the air. They perch in a tree, fly out to grab an insect, then fly back to the perch. The nest holes are becoming less frequently used, as Lewis's woodpeckers are declining in number.

Northern rough-winged swallows are in pairs now and making nests in the riverbank. You can hear juniper titmice (*Baeolophus ridgwayi*) and ruby-crowned kinglets (*Regulus calendula*) singing near the office. Various warbler species pass through while migrating north. The first hummingbirds appear. Rufous hummingbirds (*Selasphorus rufus*) pass through to breed farther north. They winter in Mexico, but they breed as far north as Alaska. Their migration can cover up to 4,000 miles (6,437 kilometers). Broad-tailed hummingbirds (*Selasphorus platycercus*) and black-chinned hummingbirds (*Archilochus alexandri*) nest here at the refuge. Different hummingbird species can hybridize. Yet, hummingbirds are very aggressive to each other around a feeder. The Aztec god of war, Huitzilopochtli, is portrayed as a hummingbird, and his mother Coatlicue is said to have conceived him after keeping a ball of hummingbird feathers in her bosom.

A few white-crowned sparrows (*Zonotrichia leucophrys*) begin to nest here, but most are just passing through as they fly to the north. Some of those migrants spent the winter here. The white-crowned sparrow song says *I-gotta-go WEE WEE now*. You can now see large groups of chipping sparrows. Black-headed grosbeaks (*Pheucticus melanocephalus*), Cassin's sparrows (*Peucaea cassinii*), Cassin's kingbirds (*Tyrannus vociferans*), and western kingbirds (*Tyrannus verticalis*) arrive. Avocets (*Recurvirostra americana*) are back as well. Long-billed curlews are in pairs but are not sitting on a nest yet. With the return of other birds, Accipiters like the sharp-shinned hawk (*Accipiter striatus*) and the Cooper's hawk (*Accipiter cooperii*) return to prey on the smaller birds.

Figure 5.10. Black-headed grosbeak
Photo: Dean Biggins

A burrowing owl (*Athene cunicularia*) occasionally stops at the refuge while heading north to nest on prairie dog colonies, although a few will nest on the neighboring Fort Union Ranch. Long-billed curlews show nesting behaviors, such as rubbing their chests with their bills, around this time in late April. You can hear turkeys gobbling. Starlings (*Sturnus vulgaris*) are nesting in cavities, particularly in cottonwood trees along the river. Bluebirds also nest in cavities, mostly in cottonwood trees, and many outside of the refuge try to provide nest boxes to help them reproduce. Bluebird nest boxes must have a hole large enough for a bluebird, but too small for a starling, or the starlings will exploit the box to the detriment of the bluebirds. Because the starling is an exotic species, conservationists are concerned about the effect they have on native species.

Starlings and house sparrows (*Passer domesticus*) were introduced to the United States from Europe.[88] The house sparrow was first introduced to New York City in 1852 to control a moth outbreak. Then in 1890 and 1891, one hundred starlings were introduced to Central Park in New York City by a Shakespearian group who wanted the birds of Shakespeare in North America.[89] By the 1950s, starlings had reached the West Coast. Both species are now widely spread throughout North America.

Turkeys can still be heard gobbling, and you can see males strutting with their tail feathers fanned. This is their peak time for breeding. Females will lay about ten eggs over the next week to ten days, but they won't start incubating until all eggs are laid. That ensures hatching at the same time. She will incubate the eggs for twenty-eight days. Raven eggs should be hatching around this time. Red-tailed hawks, ferruginous hawks, and Swainson's hawks are still sitting on eggs. Robins begin nesting, and they may have several broods in a season (April to June). They incubate their eggs for about twelve days.

Figure 5.11. Turkey
Photo: Dean Biggins

Mammals

Black bears come out of hibernation at the refuge in late April. Males emerge first. Black bears are currently the largest carnivore in northern New Mexico, although grizzly bears (*Ursus arctos*) were here at one time. The last one was killed in 1931 near Silver City, New Mexico, but reports of sightings continued into the 1950s. Black bears can be black, brown, or blond. That may confuse some people who report them as grizzlies. Grizzly bears have a dished face and a hump on their shoulders. Black bears have neither.

Black bears are omnivores, eating both plant and animal material.[90] The size of the home range depends on the productivity of the habitat. Males will have larger home ranges that overlap a few females. The new cubs may weigh about 7 pounds (3 kilograms) when first seeing the world outside of their den. They will continue to nurse until September and gain weight rapidly. They can hope for a life span of ten years, but Tom Beck reported that humans caused 85 percent of all adult and sub-adult bear mortality in Colorado. Without data in New Mexico, our situation is most likely similar to results in Colorado. When humans feed bears, it leads to more frequent contact and the death of the bear. Hunting also increases mortality. There is no evidence of compensatory reproductive success in areas where black bears are hunted.

Plants

Both white and purple fleabane (*Erigeron* spp.), verbena (*Verbena* spp.), cutleaf daisy (*Erigoron compositus*), stemless goldflower (*Hymenoxys acaulis*), fringed gromwell (*Lithospermum incisum*), Southwestern paintbrush (*Castilleja integra*), and locoweed start blooming, but it is still early for them. We have two species of locoweed on the refuge, the silver locoweed (*O. sericea*) and the purple locoweed (*O. lambertii*). You can separate the genus *Oxytropis* from the genus *Astragulus* because the *Oxytropis* has a beaked keel in the flower, but *Astragulus* doesn't. *Astragulus* is commonly called milkvetch. The first cholla (*Opuntia* spp.) blooms appear in late April and will continue blooming into the summer. We had only 0.2 inches (0.5 centimeters) of precipitation up to this point in 2018; early 2019 was wetter, and we had snow over the winter. For gardeners, peach and apricot blossoms are blooming, but a late freeze would kill the blooms.

Figure 5.12. Purple locoweed
Photo: Dean Biggins

Figure 5.13. Southwestern paintbrush
Photo: Brian Miller

Cheatgrass, an exotic species, is green and growing seed. Cheatgrass was introduced from Europe in the late 1800s.[91] That was a time of overgrazing by cattle, and of drought, so grasslands were in poor condition. Cheatgrass was thus able to establish and spread. The grass is easily identifiable because of its drooping seed heads. Those seeds are sharp and stick in the mouths of ungulates. The only time it is palatable is in the first week or so after it emerges in spring. An area overtaken by cheatgrass greatly reduces biodiversity of flora and fauna.

Its seeds germinate in the fall, and the roots grow all winter. Most native grasses have no root growth during the cold months. Because of that winter root growth, cheatgrass can out-compete native plants for water and nutrients come spring. The plant drops its seed and dries by mid-June, which can produce an early fire. While it has dropped seed in June, native plants haven't done so yet. Thus, the changed fire cycle favors cheatgrass and reduces reproduction of native vegetation. There is also some evidence that cheatgrass can change soil chemistry.[92] If cheatgrass is present only in patches, one can collect the seeds and plant cool season grasses like native western wheat grass (*Pascopyrum smithii*) and bottlebrush squirrel-tail (*Elymus elymoides*) to compete with cheatgrass. For small plots at the refuge, we mow with a mower that collects the cuttings in a bag. This grabs the seed before it falls. Seed collection needs to be done for several years because cheatgrass seeds can maintain viability for three to five years.

Figure 5.14. Cheatgrass
Artwork: Brian Miller

You will see cottonwood racemes hanging on the trees—green from the female trees and red from the male trees, giving male trees a reddish tint. Cottonwoods, black willows (*Salix gooddingii*), woods rose (*Rosa woodsii*), chokecherry (*Prunus virginiana*), coyote willow (*Salix exigua*), and peachleaf willow (*Salix amygdaloides*) start to sprout leaves. Golden currants (*Ribes aureum*) are fully leafed-out and have yellow flowers. Silver locoweed and purple locoweed are in full bloom. You can see prairie evening primrose (*Oenothera albicaulis*) flowers blooming. They have a white flower that turns pink with age. The wild plum (*Prunus americana*) is blooming. Pause and stand next to a plum tree to inhale the fragrance. You won't regret taking the time. Make a note of where the plum bushes are so you can return in July to collect the plums for fruit, wine, or jelly. Apple trees (*Malus domestica*) are also blooming.

Insects

Green-bottle flies (*Lucilia* spp.) appear, and bees and wasps buzz about. There are a few mosquitoes in the air. The males have plumose antennae and eat nectar. They are a significant pollinator. Mosquitos are vilified by humans, but they serve an important ecological function as pollinators and prey. Only the females draw blood, which is necessary for egg laying. The duskywing skippers are pretty much gone. Red-shanked grasshoppers are still common, and they provide a good meal for long-billed curlews, as do instars of other grasshopper species. Long-billed curlews have a beak that can be 6 to 9 inches (15 to 23 centimeters) long. They are very adept at using it and can catch grasshoppers in mid-flight. The grasshoppers don't stay in the bill for long. They go down the throat quickly. The bill is one way to determine the sex of a long-billed curlew. Females have a longer bill than males, and females are also slightly larger in size.

Each spring, queen bumblebees (family: Apidae, genus: *Bombus*) and social wasps (family: Vespidae) start to build nests. The nest only lasts for a single season. The first brood from the queen is made of workers, who then carry out the construction while the queen continues to lay eggs. By summer, the Vespidae nests have larger numbers, and the wasps become more aggressive when someone is near the nest. You can easily be stung numerous times. Only females in the order Hymenoptera sting because the stinger is a modified ovipositor.[93] Both Vespidae and bumblebees have warning coloration, which other insects mimic for protection.

Bumblebees are native and some of our most important pollinators. They have been declining over the last century. Factors include habitat loss for agriculture and development, importing non-native bees for commercial uses, pesticide use, and exotic pathogens. A recent study also blames climate change because bumblebees have limited adaptability to increasing heat.[94] In areas that have been warming, you are 50 percent less likely to see bumblebees than you were forty-five years ago. You can help by planting native flowers, not using pesticides, and enhancing shade in areas for cooler temperatures.

I can't let April pass without a nod to Earth Day, which began in 1970 and has been celebrated each year by hundreds of thousands. In 1970, I was a senior at the University of Illinois. At that time, Illinois had a Republican senator named Ralph Smith. He had been appointed to the Senate by the Republican governor in late 1969 after Everett Dirksen's death. Smith was part of the group who, like childhood imaginations of monsters, saw a communist hiding inside each closet and under each bed. He proclaimed that the 1970 Earth Day was a communist plot because April 22 was Vladimir Lenin's birthday. It is also my birthday (in 1948). We weren't monsters, but to Smith—well, as Joseph Heller wrote about Yossarian in *Catch-22*, we didn't have sufficient reverence for obsolete tradition and excessive authority. After being senator for a year, Smith lost his seat to Adlai Stevenson III in a 1970 special election and became a historical footnote.

CHAPTER 6

May

May 1 to May 10

Birds

Lark buntings (*Calamospiza melanocorys*) and indigo buntings (*Passerina cyanea*) arrive in early May. Lazuli buntings (*Passerina amoena*) pass through on their journey to breeding grounds in the northern United States and southern Canada. Many nesting bird species are back and increasing in number. Kestrels (*Falco sparverius*) incubate their eggs in tree cavities in early May. They will incubate for a month, then the young will fledge about a month after hatching.

Long-billed curlews are sitting on their eggs and will do so for twenty-nine days. The female curlew incubates during the day and the male at night. The sandy brown color of the birds makes their nests—a simple scrape with feathers in a bowl on the ground—hard to see when the bird is incubating because the bird blends into the grass. To find a nest, you should be up early, or look at sundown, to see the male and female switch incubating roles. Not much is known about long-billed curlews in New Mexico. Jay Carlisle is a long-billed curlew specialist with the International Bird Observatory and Boise State University who has been leading a study of long-billed curlews in the north since 2009. In 2019, he, Erin Duvuvuei (NM Game and Fish), and Kelli Stone (USFWS) expanded the study to northern New Mexico with Rio Mora NWR as a base.[95] They placed solar-powered GPS units (each unit weighs 0.4 ounces or 11 grams) to follow the birds to their winter range. Coronavirus meant that no one could put more telemetry tags on curlews in 2020.

Figure 6.1. Long-billed curlew

Photo: Luis Ramirez

The curlews are calling less at this time in May, but they are still raucous when ravens approach. Ravens will eat the eggs of long-billed curlews and geese. After curlew chicks hatch, ravens will try to grab an unattended chick with their strong beak and fly away to eat it. For taking eggs and young, ravens often work in pairs. They are highly intelligent birds. While this sounds cruel, predation is an important driver of ecosystem function and stability.

Plants

Yellow chocolate flowers (*Berlandiera lyrata*), yellow stemless goldflower (*Hymenoxys acaulis*), white fleabane (*Erigeron* spp.), red southwestern paintbrush, yellow fringed gromwell, silver locoweed, purple locoweed, and pink Santa Fe phlox (*Phlox nana*) are in bloom on the grasslands.

Take a moment to smell a chocolate flower. You can't miss them. They have eight yellow petals with a deep maroon center. The leaves are oblong, velvety, and deeply lobed, and the leaf somewhat resembles a dandelion leaf. You will be surprised how much it smells like fine chocolate. Yellow sagebrush buttercups (*Ranunculus glaberrimus*) grow in riparian areas. Cottonwoods, willows (*Salix* spp.), boxelder, and oak (*Quercus* spp.) start to leaf in early May. Cottonwood females leaf faster than males, and their racemes are green, whereas the red racemes give the males a reddish hue until they produce green leaves. Wild plum fragrance is still in the air. Enjoy the scent.

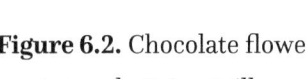

Figure 6.2. Chocolate flower
Artwork: Brian Miller

Insects

In early May, you will see tiger swallowtails (*Papilio multicaudata*), here commonly the two-tailed variety, flying around the refuge. Sphinx moths (family: Sphingidae) often dart among flowers. Sphinx moths can hover like hummingbirds and are efficient feeders on a flower thanks to their long tongues—the longest of any moth or butterfly. This is another example of convergent evolution. You can hear the hum of both the sphinx moth and the hummingbird wings as they hover. Sphinx moth caterpillars vary in color, but they do have a noticeable horn on the rear end. Common checkerspots (*Burnsius communis*) appear, as do cabbage butterflies (*Pontia* spp.). Checkerspots get their name from the checkered pattern of dark spots on white wings. They are small, about an inch (2.54 centimeters) wide at wingtips. The cabbage butterflies are slightly larger than common checkerspots and are also white. Unlike checkerspots, they only have three black dots on the upper wings. They are native to Europe but were introduced into Quebec, Canada, about 150 years ago, and they quickly spread throughout North America.

Figure 6.3. Tiger swallowtail
Artwork: Brian Miller

Perhaps one of the common insects for people in town is the army cutworm moth (*Euxoa auxiliaris*). They hatch at lower elevations, and they easily enter houses. They are harmless, but inside a house people think they are a nuisance. As spring advances, they move to higher elevations, where they are an important and rich food for grizzly bears. Army cutworm moths also eat cheatgrass that produces biological deserts when it takes over an area.

You can look for dragonflies and damselflies (order: Odonata) flying, usually over water. Their beauty never disappoints. Adults spend much of their time in the air, but the nymphs live underwater. Damselflies and dragonflies are predatory in all life stages, and the adult dragonfly has the highest success rate of any predator.⁹⁶ They capture prey on 90–95 percent of their attempts. They are carnivorous, and adults capture prey during flight, eating mosquitos, flies, butterflies, and pretty much any insect that they can take. The nymph lives in water, and it has a lower jaw that can shoot forward, like opening a drawer, to capture prey. Nymphs eat aquatic larva and even tadpoles. Depending on the species of dragonfly, the nymph stage can last from months to years. When they are ready to become an adult, they come out of the water to climb a plant to metamorphosize. You can find these dried larval skins hanging on plants near the Mora River. The shed skins resemble the shed skins of katydids.

Figure 6.4. Variegated meadowhawk dragonfly
Photo: Dean Biggins

Dragonfly migration is similar to monarch butterfly migration in that it can occur over different generations.⁹⁷ The individual that moves north in the spring isn't the same one that goes south in the fall. Research on the green darner (*Anax junius*) showed that they migrate over three generations. An adult that emerged from a larva in the north migrates to the south in fall. There, it breeds and dies. The darner emerging from the southern larva stays there to breed and produce the next generation of larva. When that larva emerges as an adult in the spring, the adult dragonfly migrates north, where it breeds, deposits larva, and dies. Then the cycle begins again.

Adult dragonflies are powerful fliers, more powerful than their damselfly relatives. Damselflies are smaller and hold their wings flat against the body vertically when resting. Dragonflies hold their wings in flight position whether flying or resting. While flying, dragonflies can move up, down, left, right, forward, and backward. Both dragonflies and damselflies use direct flight, which means the muscles attach directly to the base of the wings. Most insects use indirect flight, which means the muscles attach to the thorax. Muscle contractions change the shape of the thorax to move the wings of insects with indirect flight. Perhaps an advantage of indirect flight is the ability to fold the wings. Even though direct flight is a stronger form of flying, the ability to fold the wings may help the insects avoid predators by allowing the insect to conceal itself more easily.

Dragonflies are considered one of our more primitive insects, but they are also some of our most beautiful. Paleontologists have found dragonfly-like insect fossils from three hundred million years ago. These insects had a wingspan of up to 30 inches (76 centimeters). They could manage such a large size because of a high oxygen content in the atmosphere—30 percent compared to today's 21 percent. By two hundred million years ago, dragonflies were very similar to those of today. As oxygen levels went up and down, insect size tracked those changes until about 150 million years ago—the beginning of the Cretaceous period. At that point oxygen levels went up, but insect size declined. This coincided with the evolution of birds, so the drop in size may have given insects more maneuverability to avoid predation by birds.[98]

May 11 to May 20

Birds

You can now see western tanagers (*Piranga ludoviciana*), bullock's orioles (*Icterus bullockii*), lark sparrows (*Chondestes grammacus*), green-tailed towhee (*Pipilo chlorurus*), hepatic tanagers (*Piranga flava*), and blue grosbeaks (*Passerina caerulea*) at the refuge. If you don't have a field guide for birds, buy one to look at their delightful colors. Black-headed grosbeaks return about two weeks earlier than blue grosbeaks. Indigo buntings are in pairs. The male and female indigo buntings look like different species. The male is a brilliant blue, whereas the female looks like a brown sparrow with a little blue under the tail. Tanagers are mostly a tropical species, and they are known for their bright colors. We are lucky to have the Western tanager come to the refuge. If you're in the right place at just the right time, you may spot one in the trees around the refuge. Turkeys are quieter but still gobbling. Raven young are in the nest now, and so are hatched falcons.

Figure 6.5. Western tanager
Photo: Rich Reading

Mammals

Elk and mule deer (*Odocoileus hemionus*) sprout antlers in velvet around this time of year. May antler growth is triggered by changing day length during April. Elk antlers grow at a rate of a half inch (1.27 centimeters) per day.[99] At this point in May, you will see elk antlers that are about 6 to

8 inches (15 to 20 centimeters) long. The velvet surrounding the antler contains a network of blood vessels to support their growth. In the velvet stage, antlers are cartilaginous and not yet calcified. In general, the older the animal, the bigger the finished antler. Some elk antlers may weigh up to twenty pounds (9 kilograms). Often, a mature elk will have six points on a side; a mule deer will have four. Antler growth ends in mid-August.

White-tailed deer (*Odocoileus virginianus*) are not common in the refuge, but that species is gradually moving west, often along riparian passages, so the refuge does occasionally see one. Prairie dogs, chipmunks, and ground squirrels have young in natal dens. Prairie dog colonies are on the grasslands, chipmunks live in forests or forest edges, and rock squirrels look like tree squirrels but are ground dwellers in canyons or rocky areas.[100]

Plants

Yellow western wallflower (*Erysimum capitatum*), yellow Engelmann's daisy (*Engelmannia peristenia*), purple American vetch (*Vicia americana*), yellow golden pea (*Thermopsis montana*), and white thimbleberries (*Rubus parviflorus*) are blooming. The latter two will be nearer to water than the first three, which are grassland flowers. Golden pea, also called mountain bannertail or false lupine, has a bright yellow flower resembling a lupine. Thimbleberries produce a raspberry-like fruit. There may be a few blue flax (*Linum lewisii*) flowering, but it would be earlier than normal. Engelmann's daisies have yellow petals that extend straight from the center when they first see the sun, but as the day goes on the tips of the petals droop slightly.

The fleabane is really blooming now on the grasslands. It has a long history of use for medicinal purposes as both a diuretic and an expectorant. The name came because of a mistaken belief that the odor of the plant repelled fleas.[101] The *Oxytropis* flowers are also still strong on the grasslands. The white plum blossoms are dying off, but the chokecherry blossoms are emerging. Wavyleaf oaks (*Quercus undulata*) are leafed, but they still have hanging racemes. The bigger Gambel oaks (*Quercus gambelii*) are leafing as well, but they are a little behind the wavyleaf oak. Coyote willows have racemes that stand erect instead of hanging as cottonwood and oak racemes do. The racemes of female cottonwood trees now have small, green globes that contain the white cottony fluff and seed.

Insects

Black swallowtails (*Papilio polyxenes*) appear, and grasshoppers increase in number on the grasslands in mid-May. Grasshoppers provide a rich source of prey for birds, small mammals, and reptiles and amphibians. Other insects, such as robber flies, eat them too. Humans around the world eat grasshoppers as well, although not so much in the modern United States.

Reptiles and Amphibians

Bull snakes and rattlesnakes were active now in 2019, although this was later than the previous year. Fence lizards (*Sceloporus occidentalis*), earless lizards (*Holbrookia maculate*) collared lizards (*Crotaphytus collaris*), tiger salamanders (*Ambystoma tigrinum*), and short-horned lizards (*Phrynosoma hernandesi*)—commonly called horny toads—were also emerging at this time in 2019. The short-horned lizard is flat and round, unlike other lizards, and it has many predators. If camouflage doesn't conceal it, the lizard can inflate its body to a much larger size. If that doesn't work, it shoots blood from the corner of its eyes. It can shoot the blood, which contains a toxin that repels canids, several feet to confuse a predator.[102]

Figure 6.6. Woodhouse toad
Artwork: Anabella Miller

Woodhouse toads (*Anaxyrus woodhousii*) can mate anytime between April and June. Reproduction doesn't depend on rainfall. A female lays ten to twenty thousand eggs in a single long strand that may or may not be attached to a plant.[103] The toads can eat up to two-thirds of their weight in insects in a single day, so they are important in regulating insect numbers. A friend and I once caught a Woodhouse toad and fed it seventy army cutworm moths in a day before releasing it. The toads have a toxin in glands of the skin that make a dog or cat vomit.

Figure 6.7. Greater short-horned lizard
Photo: Shantini Ramakrishnan

May 21 to May 31

The year 2018 was very dry on the Rio Mora National Wildlife Refuge. In the fourth week of May 2018, we had 0.4 inches (1 centimeter) of rain, which brought us to 0.7 inches (1.8 centimeters) for the year. By the end of May 2019, we had almost 6 inches (15 centimeters) of rain. The amount and timing of rain in the area can be highly variable. Unfortunately, the refuge had little rain for the rest of 2019.

Birds

Kingfishers (*Megaceryle alcyon*), yellow warblers (*Setophaga petechia*), yellow-rumped warblers (*Setophaga coronata*), and yellow-breasted chats (*Icteria virens*) are back. The yellow warbler has an easily identifiable song: *sweet, sweet, sweet, I'm-so-sweet*. Yellow warblers and yellow-rumped warblers are the refuge's most common warblers, although we also have Virginia's warblers (*Leiothlypis virginiae*) and a few orange-crowned warblers (*Vermivora celata*), American redstarts (*Setophaga ruticilla*), and common yellowthroats (*Geothlypis trichas*). The arrival of warblers coincides with the hatching of goose eggs. The young goslings hatch and head directly to the water. The yellow-breasted chat was a warbler until it was recently moved to be the only member of its own family: Icteriidae. Its "song" is a series of chucks, whistles, and laughs, and the sequence varies from one call to the next.

Flycatchers (*Empidonax* spp.) and vireos return from their winter range. Lesser goldfinches (*Spinus psaltria*) and cliff swallows (*Petrochelidon pyrrhonota*) also return to the refuge at the end of May. Those two species seem to be the last of the summer residents to return for nesting, although lesser goldfinches can be seen around feeders in town during March. Cliff swallows immediately start repairing their mud nests when they arrive. If you see a cliff swallow flying in a straight line with a beak full of mud, there is water in the opposite direction. They build their nests on the cliffs near the Mora River, so keep a look out there.

Many migrating songbirds are nesting now, including house wrens (*Troglodytes aedon*), canyon wrens (*Catherpes mexicanus*), juncos, flickers, and Say's phoebes. When the eggs hatch, the parents busy themselves bringing insects to the chicks. Wrens are small birds with very loud songs. The house wren has a bubbly warble lasting several seconds. If you listen closely, you will hear *tsi, tsi, tsi, tsi, oodle, oddle, oodle*. The canyon wren has a lonely descending *dee-ah, dee-ah, dee-ah, dah, dah, dah* song.

Hummingbirds will dive-bomb any intruder who nears their nest. Hummingbird nests are very small, maybe 2 inches (5 centimeters) across, and are shaped like a cup. They are typically located in the fork of a branch.[104] The nest is made of leaves and twigs, and it is held together with spider web. Thus, the nest is elastic and can stretch as the hatchlings grow. Hummingbird eggs are about the size of a kidney bean, and there are usually two eggs in a nest. The eggs hatch in about two weeks, and the young fledge three to four weeks later. I was lucky enough years ago to monitor a garnet-throated hummingbird (*Lamprolaima rhami*) nest from egg to fledging in central Mexico. When the young were ready to fledge, they were too large to fit comfortably in the nest. They overflowed out of the top of the nest cup with almost the upper half of the young above the lip of the cup. Hummingbirds have long tongues that they use to reach nectar at the base of flowers.[105] Small tubes in the tongue carry nectar up by capillary action, so no sucking is needed. But you may be surprised to learn that hummingbirds also hawk insects on the wing or glean them from leaves and spider webs.

Male meadowlarks sing loudly and often while the female sits on a ground nest in the grass. The nest is almost invisible, and the female will not flush until you are one step away. Killdeer are sitting on nests as well, and they put on a convincing broken-wing display to lead you away from the nest location. The first long-billed curlews hatch after incubating for twenty-nine days. They leave the nest on the second day after hatching, and by one week of age they are slightly smaller than a meadowlark with a beak that is 1 inch (2.54 centimeters) long. Turkey polts hatch in late May as well.

Mammals

The first pronghorn (*Antilocapra americana*) fawns and elk fawns come in the last week of May. Most fawns will come around June 1, and the females now have distended bellies. For more on pronghorn birth, see Chapter 7, June 1 to June 10, Mammals. Coyotes, gray foxes, and swift foxes have pups in a den.[106]

Plants

With longer days, the night temperatures become warmer, which will soon spur warm-season grasses to grow. Scarlet globemallow (*Sphaeralcea coccinea*), yellow Colorado rubber plant (*Hymenoxys richardsonii*), red claret cup cactus (*Echinocereus triglochidiatus*), yellow barrel cactus flower (*Ferocactus* spp.), and yellow sweet fennel (*Foeniculum vulgare*) are blooming. The leaves of scarlet globemallow can be crushed and made into a poultice for relieving skin irritations.[107] The bulbs of wavyleaf thistle (*Cirsium undulatum*) are now a half inch (1.27 centimeters) across, and peachleaf willows start to shed their cottony seed.

Figure 6.8. Claret cup cactus

Photo: Mary Miller

CHAPTER 7

June

June 1 to June 10

Mammals

Pronghorn, bison, and elk give birth, swamping the system to avoid predation.[108] When ungulates give birth at roughly the same time, there are too many young for predators to kill and eat all of them. Both seasonal fluctuation in resource dynamics and changes in resource availability influence the degree of birth synchrony within wild ungulate populations. Thus, climate change could affect the birth synchrony of ungulate species in the future. For now, look for young ungulates on the grasslands when roses are blooming, as they do so around the first of June.

Ungulate young are precocious. They are born with their eyes open and can stand quickly after birth. They are weak, however, so for the first two weeks, they bed down alone and lay still to avoid predators. Pronghorn and elk young are well camouflaged. The mother will associate with her young around sun-up and sun-down. If you see a fawn alone, it is not abandoned. Leave it alone. Because the mother licks it repeatedly, the young have little scent.[109] By walking to the young, you may leave a scent trail for a predator to follow. You can see ravens and turkey vultures grouped around afterbirth on the prairie as proof of birth.

Figure 7.1. Pronghorn
Photo: Dean Biggins

Pronghorn young are tan; bison calves are reddish; and elk young are spotted, but the spots disappear in the fall molt. Pronghorn young grow fast, and there is a solid correlation between birth order and dominance.[110] The fawns born first within each sex are dominant over the later-born fawns. Because fawns gain about a half pound a day, a fawn born six days before another of the same sex will weigh three pounds (1 kilogram) more. Even though that difference in weight disappears over time, the dominance order remains throughout life.

Pronghorn young are twins, each weighing about 8.5 pounds (4 kilograms), so the twins make up about 17 percent of the mother's weight of about 100 pounds (45 kilograms). Elk give birth to a single young weighing about 33 pounds (15 kilograms), which is about 7 percent of the mother's weight of 500 pounds (227 kilograms). Even though both species have a 250-day gestation, a pronghorn female must invest proportionately more energy into her pregnancy than a female elk. Bison have a gestation of about 285 days, and a female weighing between 750 and 1,000 pounds (340 to 454 kilograms) gives birth to a 50-pound (227 kilogram) calf. Thus, bison investment in pregnancy is similar to that of elk.

Figure 7.2. Elk calf
Photo: Dean Biggins

Figure 7.3. Bison calf
Photo Anabella Miller

An ungulate mother will protect her young. I once watched a female pronghorn standing face-to-face with a coyote, and she successfully kept the coyote away. The young can soon run, and their speed replaces camouflage as a means of survival. A four-day-old fawn can already outrun a human. Pronghorn are the second fastest land mammal, trailing only the cheetah (*Acinonyx jubatus*) in speed. Pronghorn can run nearly 60 mph. Even though they are not quite as fast as cheetahs, they can maintain their speed for a longer time than cheetahs can. Pronghorn likely evolved their speed to escape the American cheetahs (*Miracinonyx* spp.) of the Pleistocene.[111] Even though the American cheetahs are extinct, pronghorn have maintained their speed.

Pronghorn live on the grassland, so they don't have to leap over fallen trees. Thus, they never developed a good ability to jump, as did deer and elk, which makes fences a problem for pronghorn. They try to go under fences, but the sheep fences of New Mexico have a low wire that prevents passage. Wyoming and Alberta fence law mandates that the bottom wire of fences be at least 16 inches (41 centimeters) above ground and preferably 18 inches (46 centimeters). If the bottom wire of fences is 18 inches (46 centimeters) above the ground, it allows pronghorn to slip under the bottom wire. The top two wires must be separated by at least 12 inches (30 centimeters). If the top two wires are closer than 12 inches (30 centimeters), it is easy for jumping ungulates to trap a hoof between those top two wires. Hitting the top wire can pull it below the second wire with the hoof caught between the two wires. Many New Mexico fences are not wildlife friendly, and that needs to change.

Warm season, or C4, grasses are generally preferred by many ungulates. Forbs and sagebrush (*Artemisia* spp.) are preferred by pronghorn, although pronghorn also eat grasses and other prairie plants. Deer and elk eat grass and forbs but also browse, particularly in winter. As juniper increases on western grasslands, diversity and abundance of grasses declines, erosion increases, and moisture infiltration rates reduce. This has caused desertification, and millions of acres of dry grasslands have been converted to savannas throughout the western United States. Shifts from grasses to woodlands are caused by climate (drought), heavy livestock grazing that reduces grasses, changes in the fire regime, and the loss of bison.

Bison helped maintain the grasslands, as bison break piñon pines and junipers that advance from the rocky slopes to the richer grassland soils. If you look at isolated trees on the refuge grassland, you can see marks of horning by bison. Both sexes of bison break these trees, and they do it all year. Elk males break trees only during the rut, and domestic cattle don't do it at all. The saying in Kenya is that elephants make grass, and cows make trees. Similarly, the bison behavior of breaking trees on the grassland played a role in preventing the conversion of grassland into savanna.

Figure 7.4. Bison

Artwork: Anabella Miller, Mary Miller, and Brian Miller

Plants

Rain usually increases in late June, although climate change seems to be affecting rain patterns. You can see banana yucca (*Yucca baccata*), narrowleaf yucca (*Yucca intermedia*), scarlet gaura (*Gaura coccinea*), wavyleaf thistle, and Apache plume (*Fallugia paradoxa*) plants flowering in early June. Even though silver locoweed, purple locoweed, and early fleabane start to fade, there is a rainbow of color on the grassland. Verbena, storksbill, purple locoweed, a few blue flax, wavyleaf thistle, and American vetch provide the purple. Claret cup, scarlet gaura, and southwestern paintbrush give us the red. Scarlet globemallows provide the orange. Engelmann's daisy, stemless goldflower, Colorado rubber plant, fringed gromwell, sweet fennel, western wallflower, sagebrush buttercups, and chocolate flowers show off the yellow. Fleabane, apache plume, and silver locoweed bloom white. Chokecherry blooms are fading, but three-leaf sumac (*Rhus trilobata*) is blooming in a bright chartreuse. Sphinx moths are very attracted to the wavyleaf thistle flowers, so look for them among the blooms.

Cheatgrass seed turns purple and prepares to drop seed. Bottlebrush squirrel tail, western wheatgrass, needle-and-thread grass (*Hesperostipa comata*), purple three awn (*Aristida purpurea*), and Kentucky bluegrass (*Poa pratensis*) form seed-heads, as do the hay grasses, like smooth brome (*Bromus inermis*) and orchard grass (*Dactylis glomerata*). All are cool season grasses except for purple three awn.

Cottonwoods release their seeds from the female globes, and they float in the breeze. People think the cottonwood fluff causes allergies, but it is the *pollen* that is the culprit, not the female seeds. Cottonwood males release their pollen in April. Allergies of early June are often from oak or alder pollen.

Drought speeds the conversion of grasses to woodland through competition for water. Junipers have the most drought-resistant system of water movement, called cavitation, in any plant studied. They can easily out-compete grasses. Junipers have a shallow layer of fine roots that spread in a radius from the tree to about three to four times the height of the tree. The roots are less than 6 inches (15 centimeters) below the surface, and those roots take moisture that could otherwise go to grasses. When several juniper trees are close together, the overlapping halos of fine roots can completely denude the soil. If this is on a slight slope, a rivulet can form on the bare ground, which over time becomes an arroyo. If there is little rain, junipers have a tap root that can extend 150 feet (46 meters) below the surface. In addition, most piñon and juniper seedlings take root on soil without much grass cover. Good grass cover on the deeper grassland soils effectively hinders grassland colonization by the conifers. Overgrazing weakens the grasses and leaves more bare ground for tree and shrub colonization.[112]

There has been a fundamental shift from grasslands to wooded savannas in the last 125 years. Predictions for increased drought with climate change will exacerbate the problem. If fragmentation proceeds in a random fashion, when 40 percent of an original habitat type is left, the patches become disconnected. The isolated patches of that habitat-type then continue to shrink and become more distant from each other, with the invading habitat-type becoming more dominant. Loss of grasslands to expanding woodlands eventually eliminates the matrix of different habitat types in an area and thus removes ecological boundaries that allow greater diversity of species to exist across a landscape.

Insects

Long-billed curlew chicks are still hatching, and chipping sparrows are sitting on nests. Both feed on insects during summer. Juniper hairstreaks (*Callophrys gryneus*) appear and so do bee flies (*Bombylius* spp.). Tent caterpillars are now on three-leaf sumac, oak, and chokecherry. Tent caterpillars are moth larvae, belonging to the family Lasiocampidae and the genus *Malacosoma*. There are six species in North America, and in northern New Mexico we have the western tent caterpillar (*Malacosoma californicum*). When larvae emerge in May or June, they colonially build a tent to protect themselves from predators. They defoliate leaves inside of the tent. This kills the branch, but the tree usually refoliates. Because tent caterpillars have boom and bust cycles, in extreme conditions some trees may die. As the larvae age, they may move out of the tent to feed, but they return to the tent at night. Being social, a caterpillar returning from food leaves a pheromonal trail that hungry caterpillars inside the tent can follow to feed. Six to eight weeks after becoming caterpillars, they pupate, and the adult moth emerges two weeks later. They lay eggs that overwinter.

Cecropia moths (*Hyalophora cecropia*) are out in late spring and early summer. These giant silk moths are found throughout the United States and Canadian provinces, but a sighting is rare. They have no mouth or digestive system, so they only live a few weeks. During that time, they mate and lay eggs. When the eggs hatch, the caterpillars mature in fall, then spin a cocoon that hatches the next spring or early summer when temperatures warm.

June 11 to 20

Birds

Canada geese start to molt their flight feathers, and then in July, they replace all their feathers. The molt lasts about a month, and during that time the geese can't fly.[113] They pick a spot near open water during the molt. This allows easy access to food and escape. Many smaller birds are now feeding young in the nest. If a nest still has eggs, those eggs will hatch soon.

The young bald eagles are nearly full sized, but they are still in the nest. They are dark, and they won't have a white head and tail until they are five years old. The curlew chicks are about 8 inches (20 centimeters) tall with a 2-inch (5 centimeters) bill by mid-June. Nighthawks (*Chordeiles minor*) are back. They are graceful, nocturnal fliers who catch insects on the wing. They make a *peezt* call while in the air. In summer around dusk, males will fly above the trees, then suddenly dive toward the ground. When he pulls up from the dive, he adjusts his wings so that the air rushing through the feathers makes a loud whooshing sound. These are courtship dives, but they can also be aimed at territorial intruders.

Mammals

Bison are done calving by mid-June. Mule deer are now giving birth after a gestation of 205 days. The young deer are spotted like elk young, and they are well camouflaged. A female giving birth for the first time has a single fawn, but older females in good condition have twins.[114] A single fawn weighs about 9 pounds (4 kilograms), and an adult female can weigh about 150 pounds (68 kilograms). Twins can thus weigh about 12 percent of the mother's weight. While the investment of a female mule deer seems lower than that of the pronghorn, the mule deer gestation is forty-five days shorter. The young pronghorn and elk are now strong enough to join the herd and run.

Coyote and fox pups emerge from the den. Coyotes are one of the most widespread carnivores in North America. They are very adaptable. Lewis and Clark reported the first coyote they saw as being in what is now Nebraska, but the species has spread to the Atlantic Ocean. About the only way to control coyote numbers is by having wolves. Wolves and coyotes do not get along well. In some cases, wolves not only kill coyotes, but they also rip those dead coyotes apart.

In fairness, coyotes do the same to swift foxes and black-footed ferrets (*Mustela nigripes*). In both cases, coyotes kill the ferrets and swift foxes, but don't eat them. Indeed, coyotes can suppress the numbers of other small carnivores. This happened when wolves were extirpated from the Greater Yellowstone Ecosystem. Without wolves, coyotes became dominant and increased in number, and that depleted the numbers of other small carnivores. The reappearance of wolves reduced coyote numbers and forced coyotes to use habitats providing protection. We studied how recolonizing wolves affected coyote numbers and distribution in Grand Teton National Park. In areas without wolves, coyotes used the marshy grasslands where vole (*Microtus* spp.) numbers could be above 50 voles per acre (123 voles per hectare). In areas with wolves, coyotes abandoned those good hunting areas for pine forests and sagebrush, where rodent numbers were 5 to 10 per acre (12 to 25 rodents per hectare), but there was better concealment from wolves.[115]

Most of the coyote diet is rodents and lagomorphs, up to 80 percent of it in some studies.[116] They also take pronghorn fawns in the spring. An adult coyote needs about 21 ounces (600 grams) of meat a day, and a lactating female needs about 32 ounces (900 grams).[117] Springtime fawns help those lactating females. After the young are weaned, their diet returns to small prey. The presence of a top predator increases diversity in the community, particularly among small carnivores. When wolves kept coyote numbers in-check, smaller carnivores thrived. Joel Berger found that pronghorn fawns survived better in areas with wolves because of their effect on coyotes.

In mid- and late June, female black bears chase off their adolescent young. The cubs are born during hibernation and den with the mother over the next winter. At one and a half years of age, the young will be on their own. The adult female will come into estrus, and if young are still with her, a male coming to breed may kill the yearlings. Bears have induced ovulation, which means that the act of copulation causes the female to ovulate. Bears also utilize the strategy of delayed implantation. The blastocyst floats freely and is dormant. It doesn't implant in the uterine wall until November or December. Thus, the female can forgo the energetic demands of pregnancy while gathering fat for winter hibernation. If she doesn't achieve sufficient fall nutrition, the blastocyst is reabsorbed at little cost to the female.[118]

Plants

Old plainsman (*Hymenopappus artemisiifolius*), poison hemlock, shrubby cinquefoil (*Dasiphora fruticosa*), kidney vetch (*Anthyllis vulneraria*), cutleaf germander (*Teucrium botrys*), yucca (*Yucca gloriosa*), salsify (*Tragopogon pratensis*), Chihuahuan flax (*Linum vernale*), mountain ninebark (*Physocarpus monogynus*), thimbleberry (*Rubus parviflorus*), spiny aster (*Chloracantha spinosa*), James penstemon (*Penstemon jamesii*), and yellow sweet clover (*Melilotus officinalis*) are in bloom. The refuge has a nice patch of blooming Canadian anemone (*Anemone canadensis*) in a shady spot near the Mora River. They are white and resemble geraniums. New Mexico locust (*Robinia neomexicana*) is in full bloom with rose-pink flowers. Wild asparagus (*Asparagus officinalis*) and boxelder trees

(*Acer negundo*) form seeds. Wooly plantain (*Plantago patagonica*) and cheatgrass is casting seed and dying. Locoweed flowers are gone by this point in June, and Buffalograss (*Bouteloua dactyloides*) has gone to seed.

Figure 7.5. James penstemon
Photo: Dean Biggins

Insects

When we monitored grasshoppers on the high grasslands in 2018 and 2019, snakeweed grasshoppers (*Hesperotettix viridis*) were abundant. Snakeweed grasshoppers mainly eat snakeweed, which is a plant that ranchers do not like because when a cow eats it, she may abort her fetus. Pronghorn, however, can eat snakeweed with no issue. In fact, in some areas, up to 30 percent of the pronghorn diet is snakeweed. On monitoring transects in June 2018 and 2019, we recorded about ten thousand grasshoppers per acre of the upland grassland, and 60 percent of that number were snakeweed grasshoppers. In 2020, the flabellate grasshopper (*Melanoplus occidentalis*) dominated. Again, there were about ten thousand grasshoppers per acre, with 60 percent being the flabellate grasshoppers.

Robber flies (*Asilidae*) hunt on the refuge in open, sunny areas. They are aggressive predators who take other insects on the wing. If you handle them, they can deliver a painful bite. We commonly see a robber fly who has captured a grasshopper, which is a nutritious meal. Both Native Americans and later settlers ate grasshoppers fried or roasted over coals. The practice has declined in modern times, but in Mexico you can still buy grasshoppers by the bagful in the markets. They taste good, perhaps a little like shrimp. I prefer them mixed with scrambled eggs. Grasshoppers are rich in antioxidants, but some people have an allergic reaction to them.

When a robber fly is perched with a captured prey item, you can get very close for pictures. When other insects become scarce, robber flies can become cannibals. Darkling beetles (*Tenebrionidae*) walk around throughout the refuge. If they are disturbed, they raise their hind end and emit a foul smell. In primary school, my daughters liked to pick up the darkling beetles to get the smell on their fingers. They did the same with garter snakes.

If you are lucky, you may see a termite swarm on a warm day after a rain. The swarmers are males and females capable of reproducing, and they fly in dense numbers. After flying away from the nest, the female attracts a male with pheromones. They then break off their wings and mate to start a new colony. Termites are social insects with workers and soldiers, but neither of those castes have wings.[119]

June 21 to June 30

Summer Solstice

June 21 (plus or minus a day) is the summer solstice. We discussed the tilt of the Earth on its axis in the section on the spring solstice. The tilt of the Earth is 23.5 degrees, and at the summer solstice the sun shines directly on 23.5 degrees north, meaning it brings the longest day of the year for northern New Mexico. If you live in the southern hemisphere, however, this solstice means that you are in winter and seeing the shortest day of the year. Although the tilt of the Earth can change, and occasionally does, the changes are slight. Earth has a large moon that stabilizes the tilt to minor shifts. Planets with small moons, like Mars, can shift their tilt to a higher degree.

Birds

In 2019, the two bald eagle chicks at the refuge fledged the nest on June 21. It was very much the same in 2020. One day, they are sitting on the edge of the nest. The next day, both young are gone, and the two adults are perched on limbs close to the nest. The day after that, the nest and tree are empty. Most bald eagles of New Mexico come down from the north for the winter. The few bald eagles that nest here—only about a half dozen—are all in the northeastern part of the state.

Female long-billed curlews start to fly south, leaving the chicks with the male. The chicks grow fast and are approaching adult size. In about three or four more weeks, the male and young will also fly to their wintering area.

Figure 7.6. Long-billed curlew
Artwork: Mary Miller

Mammals

Prairie dog pups emerge from their underground dens in late June. Elk, deer, pronghorn, and bison are visible with their young, and the winter coats of bison are gone, and they now look quite sleek.

Young desert cottontail rabbits (*Sylvilagus audubonii*) are on their own and are perhaps half the size of an adult. They try to stay motionless and avoid being seen if you approach. Cottontails like prairie dog colonies because the burrows provide excellent cover. They also use crevices, rocks, and shrubs for cover. If freezing fails, they escape by running. We had one near the office on the refuge that would run into a rainspout if scared.

Figure 7.7. Desert cottontail
Photo: Dean Biggins

Black-tailed jackrabbits (*Lepus californicus*) are in the same order as cottontails (Lagomorpha), but they're larger with longer legs. They have a black stripe on their tail that extends to the rump. Another difference is that cottontails are born hairless with eyes closed (altricial birth), whereas jackrabbits are born fully haired, eyes open, and ready to move around (precocial birth).

A host of predators like to eat both. During the Dust Bowl, people organized mass jackrabbit hunts because they thought jackrabbits ate too much forage. The Lagomorph order is very old, with fossil evidence from sixty million years ago.[120] Black-tailed jackrabbits have been declining, probably because of habitat changes, overgrazing, and expanded cultivation. Cottontail numbers, however, seem stable.

Figure 7.8. Jackrabbit
Artwork: Brian Miller

Plants

Plains prickly pear (*Opuntia polyacantha*) bloom in a beautiful shade of yellow. Broadleaf milkweed (*Asclepias latifolia*) begins to show flowers, and cota (*Thelesperma megapotamicum*) begins to flower as well. Cota is native to this area and has deep importance for many in the state as a medicinal plant. It is used to relieve an upset stomach, to reduce fevers, to eliminate intestinal worms, and to reduce tooth pain.[121] It makes a very nice tea if you boil the plant for 5 or 10 minutes. You can store it in dried bundles about 5 inches (13 centimeters) long and an inch (2.54 centimeters) or less wide. Tie each bundle with another stem of cota, and you have a ready mixture for a half gallon of hot or cold tea. The yellow flowers can also be boiled to make a natural dye. From late June through July is the prime time to collect cota for tea.

The berries of three-leaf sumac are full sized and turn from green to red. If you gather enough of the red berries to fill a coffee mug, you can mash and boil them. When you have a half gallon of water with the strained sumac juice, add a can of frozen lemonade to sweeten the drink. As a caution, do not eat any plant that you can't positively identify.

Prickly pear cactus is edible and very tasty. It is a popular dish in Mexico. The fruit (or *tuna* in Spanish) has a sweet melon taste. It is rich in antioxidants and anti-inflammatory properties; thus it has shown health benefits for people with inflammation or joint problems. It also reduces cholesterol and triglycerides. The pad of prickly pear (*nopal* in Spanish) is peeled and cooked or eaten raw. Be careful to eliminate spines from both the pad and the fruit. In the field, the cactus fruit can be a reliable source of water to quench your thirst if your canteen is dry. The nipple cactus (*Mammillaria grahamii*) has a red fruit that to me is the freshest sweet taste.

Insects

Dog day cicadas (*Tibicen duryi*) and walking cicadas (*Okanagana synodica*) buzz throughout the refuge in late June. Dog day cicadas got their name because they come during the hottest days of summer and fall. They are larger than the walking cicadas, which are found in June and July.

The males make the buzzing noise by a percussion mechanism. The mechanism is a tymbal on each side of the thorax or abdomen, and the male vibrates that tymbal by muscular contraction. Dog day cicadas and walking cicadas are above ground every year, but most of their lives are underground as nymphs. The adult lays eggs on twigs, and when the eggs hatch, the nymphs fall to the ground. The nymphs then burrow into the ground and attach to a root where they suck sap. After several years of living below ground, they dig to the surface and climb a plant. They split their exoskeleton, and the adult emerges to restart the process by laying eggs on a twig.[122]

CHAPTER 8

July

July 1 to July 10

Summer Rains

Rain should have started by now, although the effects of climate change may be pushing the rainy season later into the summer. In addition, the rain has been less frequent over the last fifteen years. Several decades ago, there was light rain each afternoon in early July. Now we have a strong storm, then a dry period of ten to fifteen days before there is another one. Because strong rains tend to run off, and the hardened ground during the dry period between rains increases that run off, plants receive less moisture. The average precipitation of occasional strong rains may be the same as the average rainfall several decades ago when there were light rains every day, but there is less meaningful moisture for plants today with the changed climatic condition.

Birds

All of the female long-billed curlews have flown south, leaving the chicks with the male. Some of the long-billed curlews with GPS transmitters flew directly to the Sea of Cortez, a place where the Colorado River enters the Gulf of California.[123] They pretty much made the trip in one day.

Hummingbirds are not easily seen at this time because they still have young in the nest. When those young fledge and the migrating rufous hummingbirds arrive, it will seem like hummingbirds are at every feeder.

Figure 8.1. Cone flower
Photo: Anabella Miller

Plants

In early July at the refuge, color floods the grasslands. Flowers commonly seen in shades of yellow are Englemann's daisies, Rocky Mountain zinnias (*Zinnia grandifolia*), suckling clover (*Trifolium dubium*), golden tickweed (*Coreopsis tinctoral*), and prairie cone flower (*Ratibiba columnifora*). The prairie cone flower comes in both yellow and orange. There are pink waves of Santa Fe phlox, prairie smoke (*Geum triflorum*), and the pinkish showy milkweed (*Asclepias speciosa*). Scarlet penstemon (*Penstamon barbatus*) shows off its bright red color. Shades of purple abound in purple geranium (*Geranium caespitosum*), spike verbena (*Verbena macdougalii*), scurf pea (*Psoralidium teniuflorum*), silver nightshade (*Solanum elaegnifolium*), prairie clover (*Dalea purpea*), and cliff primrose (*Pimula rusbyi*). You will see white flowers from prairie clover (*Trifolium repens*), valerian (*Valeriana officinalis*), paper daisies (*Rhodanthe* spp.), and antelope milkweed (*Asclepis asperula*). The latter is actually a greenish white. Chocolate flowers, with their yellow petals, and fleabane, which come in both white and purple, are vanishing. Thimbleberries have green berries, but they are not yet ripe.

Insects

Carolina locusts (*Dissosteira carolina*) are a more common appearance in early July. They are large and tend to be near water. Plains lubbers (*Brachystola magna*) are present as instars. They will be adults soon. Lubbers can't fly, because they have small vestigial wings. They are the size of a human thumb, but their wings are only a quarter of an inch (0.6 centimeters) long. Both lubbers and Carolina locusts are large Orthopterans and valuable prey for many birds and mammals. Neither is a threat to grasslands. In fact, the Carolina locust eats weeds, including cheatgrass.

Figure 8.2. Plains lubber
Photo: Chris Wemmer

The first monarch butterflies (*Danaus plexippus*) and queen butterflies (*Danaus gilippus*) are scattered about the refuge in the milkweed.

Monarch butterflies will lay one egg under the leaf of a milkweed plant.[124] They lay one egg per leaf because if there are more, the first caterpillar to hatch will eat the other eggs. The egg hatches in slightly less than a week, then it spends ten days as a caterpillar and another ten days as a chrysalis. Milkweed is the only food for the caterpillar. Cardiac glycoside is a toxin in the milkweed leaf, but monarchs have a genetic mutation making them one hundred to two hundred times less susceptible to that toxin than other organisms. By ingesting the toxin, the caterpillar and butterfly are then protected from predators. Their orange and black coloring warns predators to stay away. Using bright colors as a warning is called aposematic coloration. The female monarch can assess the level of glycoside in a plant before laying its eggs. Too little toxin would fail to give protection, and too much may kill the caterpillar.[125] The monarch in Figure 8.3 is a male. Males have a small black dot on the inside line of the hind wing.

Figure 8.3. Male monarch butterfly
Photo: Anabella Miller

The midwestern monarchs fly north for three generations.[126] The transition from egg to butterfly to egg takes about five weeks. The fourth generation lives about seven months, and in August it will fly 2,000 to 3,000 thousand miles (3,219 to 4,828 kilometers) to one of twelve mountaintops in Central Mexico. This is an incredible journey because that generation has never been to the winter ground. The mountaintop habitat is an oyamel fir (*Abies religiosa*) forest with the correct temperature range. During winter, monarchs are in reproductive diapause, but in spring, they fly north and lay the first generation in the Gulf states. The cycle then starts again.

Oyamel fir is a relict forest of wetter and cooler times that is now restricted to high elevation in central Mexico. Almost all midwestern monarchs spend the winter on the twelve mountaintops. This restricted winter area makes the population very vulnerable. Monarchs benefit from the cool temperatures because they can be physiologically active, but the cool weather means they expend less energy in that activity, conserving it for the spring flight. Oyamel fir is mostly found at the tops of mountains now, and as temperatures warm, it will likely disappear. Illegal logging also takes a toll. The monarch winter habitat needs a closed canopy to control conditions for monarch survival, so even thinning trees is harmful.[127] The trees are also cleared to grow avocados, and drug cartels are now entering the avocado market.[128] Of course, the trees cut to plant avocados can be sold on the illegal market. Even though it is illegal to cut the trees, you can see semitrucks on the highways with loads of timber.

We visited the Piedra Herrada monarch winter area in Mexico, and words can't really describe it. Trees turn orange with butterflies. Monarchs west of the Continental Divide winter in California, and that population has declined by almost 90 percent. Not much is known about monarchs in New Mexico.

Monarchs west of the Rocky Mountains winter along the Pacific coast of California near San Diego. They use eucalyptus trees (*Eucalyptus* genus but with 650 species), pines, and cypresses that provide a similar environment to the Oyamel fir forests on Central Mexico. There is much to learn about monarch butterflies in New Mexico. Some may go to either winter site.

Contrary to popular belief, monarchs are not good pollinators. Their legs are too long to allow contact with pollen sacs. According to Anurag Agrawal in his fantastic book *Monarchs and Milkweed*, the monarchs need the milkweed, but it doesn't need the monarchs.[129] Even though some agencies highlight monarchs as pollinators in education programs, bees (order: Hymenoptera) are the best insect pollinators.

July 11 to July 20

Birds

All birds are in the last stages of incubating, feeding young, or have already fledged their young by mid-July. Cliff swallow young fledge their mud houses on cliff faces by the Mora River, and the adults and fledglings leave their colonies to begin migration. Young ravens are independent. In the morning, a raven family spreads across the grassland evenly spaced. They simply walk around to catch grasshoppers. They have molted their feathers, and the grasshoppers provide energy to replenish the cost of molting.

Mammals

Bison males start to posture to each other in mid-July. Posturing behavior shows dominance and includes the male bellowing, sniffing the female's genitals, and wallowing violently to kick up dust. A raised tail is a threat sign. The rut will come in a couple of weeks.

Badgers breed in July and August. During the breeding season, male badgers range widely, and they can pay a price when crossing roads. Like the bears mentioned earlier, they utilize the strategy of delayed implantation.[130] About a hundred species of mammals use this reproductive strategy. They also have induced ovulation, where the act of copulation stimulates ovulation. This is advantageous to species who are solitary and widely spaced. It provides some assurance that the copulation will be successful, as opposed to a strategy where the female comes into estrus and hopes that a male finds her in time. Because of delayed implantation in badgers, the blastocyst doesn't implant in the uterine wall until spring. Young are born in April. Badgers are very tough, and therefore, they have few enemies. I once had a badger chase me, and I had to jump on the hood of a truck to escape. Prairie dog pups are easy prey for badgers.

Figure 8.4. Badger
Photo: Dean Biggins

Another predator of grasshoppers and insects around the refuge is the northern grasshopper mouse (*Onychomys leucogaster*). While most rodents eat seeds or vegetation, grasshopper mice are predatory, sometimes even eating other small mammals. Their body is about 5 inches (13 centimeters) long and accompanied by a short tail. Their long, sharp incisors aid their predatory abilities. They live in burrows during the day, either digging the burrows themselves or using prairie dog burrows. At night, they emerge to hunt. Before they begin, they howl to announce to other grasshopper mice that they are there. There was once a Peter Sellers movie called *A Mouse That Roared*. This is the only mouse that can do that. They are a unique and handsome rodent, although rarely seen. When biologists live-trap rodents, they usually use peanut butter and oats, but that bait doesn't appeal to the grasshopper mouse. Insects are a more reliable bait. Grasshopper mice are hard to trap, but they may enter a box trap if it contains insects.

Plants

Pineappleweed (*Matricaria discoidea*), coyote gourd (*Cucurbita palmata*), ox-eye sunflower (*Heliopsis helianthoides*), false boneset (*Brickellia eupatorioides*), silverweed cinquefoil (*Potentilla anserine*) all start to bloom. Pineappleweed has a tiny yellow flower, and when you step on the plant, it gives a pleasant odor similar to the smell of pineapple. Coyote gourd, ox-eye sunflower, and silverweed cinquefoil all have yellow flowers as well. Ox-eye sunflower resembles a sunflower but is really not part of the sunflower genus. False boneset has a white flower and pollinators like butterflies and bees are attracted to the plant. Silver bluestem (*Bothriochloa laguroides*) form seed heads in mid-July. Silver bluestem is a warm season bunch grass that is drought tolerant. Wildflower colors start to diminish by this time in the summer.

Insects

In mid-July, flies become very active. If you walk outside, you will soon have a small swarm that circles your head. Deer flies and horse flies (family: Tabanidae) can deliver painful bites with their sharp mouthparts. They can be found from late June until September, but July is the main month. Their life span is around thirty to sixty days. The females feed on the blood of mammals, which they need to reproduce. Males eat pollen and thus pollinate plants. A horsefly can take 1 cubic centimeter (cc) of blood from livestock. Humans quickly swat at the bite, so we are not a good source of blood. That said, when a female is forced to move between mammals for a meal, she can transmit diseases like anthrax and tularemia. Predators of Tabanidae include wasps, dragonflies, and insectivorous birds. They lay their eggs in wetlands, so humans can't control the flies with insecticide sprays. Unfortunately, DEET applied to the body doesn't repel the flies, so the best protection is to wear long sleeves.

July 21 to July 31

Birds

Lark sparrows, horned larks, cliff swallows, meadowlarks, and Vesper sparrows fly in large groups and hang out along the road by late July. The young ones are not smart about vehicles. They fly in the same direction that the car is going instead of flying to the side. If you drive faster than 40 mph, you are bound to hit some birds along the state highway to Rio Mora NWR. Long-billed curlews have left the area, as have mountain plovers. There are lots of hummingbirds now as young have fledged the nest, and the rufous hummingbirds travel through from their northern nesting areas in Canada and Alaska to the southern winter range in Mexico. This journey can cover almost 4,000 miles (6,437 kilometers). While rufous hummingbirds are the most aggressive at protecting a feeder, they also have a longer migration than other hummingbird species, so they need the energy. Canada geese have completed their molt and can now fly again.

Plants

Yarrow (*Achillea millefolium*) and Rocky Mountain bee plant (*Cleome serrulata*) start to bloom. Plains pagoda (*Monarda pectinata*) has become common. Both the purple Rocky Mountain bee plant and the white plains pagoda are attractive to pollinators. Blue grama (*Bouteloua gracilis*), side-oats grama (*Bouteloua curtipendula*), wolf-tail (*Lycurus phleoides*), and vine mesquite (*Panicum obtusum*) form seed heads.

Insects

Blister beetles (*Epicauta* spp.) appear from mid-summer through late summer. They can appear suddenly as a large swarm, then disappear just as rapidly. We saw this on the refuge. One day, black blister beetles (*Epicauta subglabra*) arrived and fed on the tops of kochia plants (*Kochia scoparia*). Kochia is an exotic plant from Europe, and it is extremely drought resistant. Thus, it does well in this part of New Mexico. When the black blister beetles arrived, there were multiple individuals on each plant in an area with abundant kochia. The next day, the beetles were nowhere to be seen. Blister beetles feed mainly on pollen, but they can also eat flower parts and leaves.

Blister beetles lay eggs in the soil. According to David Lightfoot, a grasshopper biologist from the Museum of Southwestern Biology, when the blister beetle eggs hatch, the prelarvae eat grasshopper eggs that have been deposited about an inch (2.54 centimeters) or so below ground. The prelarvae have long legs and are active in search of grasshopper egg sacs. Thus, blister beetle numbers can coincide with grasshopper fluctuations. David said how long the larvae persist below ground is not well-known. This could be a good study for a graduate student. The prelarvae also eat the eggs of bees that nest below ground, like the sweat bees of Halictidae. The larvae find bee eggs by climbing onto flowers. They attach themselves to bees coming to those flowers, and the bee then brings the attached larvae back to its nest.[131]

The cantharidin that blister beetles produce is toxic and stored in their blood. The amount of the toxin varies among species, and it is more concentrated in males than it is in females. Fortunately, the black blister beetle is on the low end of the cantharidin scale. Nevertheless, when black blister beetles exude cantharidin, it can inflame your skin and sometimes cause blisters. When we lived at the Chamela-Cuixmala Biosphere Reserve in Jalisco, I woke up one morning with a burning sensation on the skin of my leg. I had several blisters, one of which was over an inch (2.54 centimeters) long. Apparently, there was a blister beetle in our bed. Thankfully, it preferred me to my wife. I never found the beetle, but our house was frequented by insects and an occasional scorpion. My blisters took about a week to heal.

CHAPTER 9

August

August 1 to August 10

Birds

Nesting activity is essentially over at the refuge in August. Our local birds are in large groups and ready for fall migration.

Mammals

Elk and mule deer antlers are now full size for the season, but they are still in velvet. Bigger animals have bigger antlers. A mature set of antlers comes at about six years of age for mule deer and at about ten years of age for elk, although it varies with genetics and level of nutrition available to the animal.[132] When antlers are full size and stop growing, the animal sheds the velvet. Because antler growth is so fast, it is somewhat of a handicap. The nutritional demands of antlers are great, but the payoff is that the dominant males can breed. That said, if antlers were only for reproduction, the advantageous thing would be to shed them right after the breeding season, yet they maintain them throughout the winter. Males shed in February or March. Winter is a vulnerable time for ungulates, so antlers may be a defense against predators.

Figure 9.1. Elk
Artwork: Brian Miller

Figure 9.2. Mule deer
Artwork: Brian Miller

Bison and pronghorn do not have antlers. They have horns. Horns are a bony projection from the skull that is surrounded by a keratinous sheath. Both sexes have them. Female pronghorn have smaller horns than males, and males also have a black cheek patch that is absent on females (Figure 9.3). The horns of a male bison have a wider spread than female horns. Unlike antlers, horns are permanent and not lost in spring. Pronghorn shed the keratinous sheath once per year. The time for shedding the pronghorn sheath is variable. The newly forming sheath will push the old sheath off of the bony core.

Figure 9.3. Male pronghorn
Artwork: Brian Miller

Bison calves turn black, and the adult males have started rutting. Males use a Flehmen response to detect pheromones and estrus in females. By curling the lip and closing the nostril, scents are passed directly to the vomeronasal organ, which is above the roof of the mouth (Figure 9.4). When a female comes into heat, the dominant male will push her to the side of the herd and guard her until he can mount and inseminate her. He will then return to the herd to seek a second female. If a female comes into season while the dominant male is guarding another female, then the male who is second in line will push that female to the side. This is a reason that having multiple bulls will increase reproductive success.

Figure 9.4. Bison Flehmen response
Photo: Anabella Miller

Female bison can reproduce at two or three years of age, but three years of age is more common on the refuge. Gestation is 285 days. Bison females also use the Flehmen response to synchronize estrus. Males are sexually mature at three years of age, but they don't have much chance of breeding until they are older.

The refuge participates in the Pueblo of Pojoaque bison program for ecological and cultural restoration. Both bison and prairie dogs are key in maintaining healthy grasslands. But in the late 1880s, for their reward, they were shot and poisoned. Culturally, bison had great significance to native people of the western plains. By 1880, bison were nearly eliminated in the West and replaced with cattle. General Sherman, who had used scorched-earth policies during the Civil War, was appointed by President Grant to lead the war against Native Americans in the western theater. In 1877, according to John Cook, a professional bison hunter in the 1870s, Texas proposed a law to protect bison. In 1906, at the age of sixty-two, Cook decided to write down his experiences from those days.

In describing the decline of bison in the 1870s, Cook wrote that General Sheridan went to the Texas legislature to lobby against that law.[133] To quote Sheridan's speech:

> These men [bison hunters] have done in the last two years and will do more in the next year, to settle the vexed Indian question, than the entire regular army has done in the last thirty years. They are destroying the Indian's commissary; and it is a well-known fact that an army losing its base of supplies is placed at great disadvantage.... For the sake of a lasting peace, let them kill, skin, and sell until the buffaloes are exterminated. Then your prairies can be covered by speckled cattle, and the festive cowboy, who follows the hunter as a second forerunner of an advanced civilization.[134]

This quote is controversial, however. Flores claims that the quote is questionable because there is no primary source and no evidence that Sheridan addressed the state legislature. Flores speculated that Cook cast it as a military necessity to make himself, and perhaps all Americans, feel better.[135]

Nevertheless, there was a strong link between Plains tribes and the bison. There were multiple effects of the bison decline. The market for hides was certainly one of them. David Smits, a historian, argued that the US Army was primarily responsible, and General Sherman stated more than once that bison eradication would force the tribes to reservations.[136] Although the market hunters were civilian, Sherman commended them, didn't enforce treaties, protected them, and sometimes gave them ammunition. Market hunters provided bison meat to forts and railroad workers, an early form of bushmeat. Colonel Dodge added, "Kill every buffalo you can. Every buffalo dead is an Indian gone."[137] In 1872, Colonel Dodge estimated that for every hide shipped to market, of which there were 1,378,359 between 1872 and 1874, there were another five bison killed.[138] Indeed, the army was willing to massacre not just the four-legged, but also the two-legged as seen at Sand Creek, Washita River, and Little

Bighorn.[139] Clearly, killing the bison was seen as a way to starve the Plains tribes into surrender, as much as they were victims of industrial society.

It is interesting that of all the ungulates facing extinction in the late 1800s, only the bison has been unable to recover significant wild numbers. If you look at only the lower forty-eight states and southern Canada, you can also include woodland caribou (*Rangifer tarandus caribou*) as a recovery failure. Overall, in Canada, about 30 percent of the woodland caribou populations are large enough to be self-sustaining. Both woodland caribou and bison are near functional extinction.

Bison played an important role in grasslands. Many ungulates, who are scattered grazers, follow a green-up in the spring migration. This is called the green wave hypothesis, or sometimes surfing the green wave. The new grass is highly nutritious. As grasses mature, nutrition declines, and scattered grazers move to a higher altitude to exploit the new green-up. Bison, however, do not have to follow the green wave in Yellowstone. Because they are aggregate grazers, their intensive grazing can increase the nutritional content of the plants in an area by as much as 50 percent. They create the conditions that they desire and fall behind the green wave. According to a recent paper in the *Proceedings of National Academy of Sciences*, aggregate grazing keeps grass in an early and nutritious stage with a high photosynthetic rate.[140]

This pattern in Yellowstone National Park is similar to the grazing lawns of the Serengeti, where the grazing lawns can increase nutritional value up to 90 percent.[141] That nutritional increase is probably higher than in Yellowstone due to the tropical warmth. Because aggregate grazing delays plant maturation and increases leaf to stem ratio, plant biomass of grazing lawns also increases biomass by 30 percent in the Serengeti even though the plants are shorter. Cattle are not aggregate grazers, and that may explain why our grasslands deteriorated when bison, and prairie dogs, were replaced by cattle. Even though the North American prairies evolved with grazing, they evolved with many aggregate grazers. Indeed, when looking at bovids from a great distance, you can usually tell bison from cattle because cattle are more scattered while bison are clumped together.

Another important grassland species is the prairie dog (*Cynomys* spp.). In 2007, we released Gunnison's prairie dogs (*Cynomys gunnisoni*) onto the Wind River Ranch, which was the predecessor to Rio Mora National Wildlife Refuge. We chose Gunnison's prairie dogs because they hibernate, and thus they are mainly active during the growing season. Furthermore, they space themselves farther from each other than other prairie dog species, so Gunnison's prairie dogs may be more suited to drought. In 2016, they were eliminated from the Rio Mora NWR by **plague bacteria** (*Yersinia pestis*). We dusted to kill fleas in the burrows every year for seven years, but apparently the fleas developed resistance to the dust over time. When plague prevention improves, we would like to re-establish a Gunnison's prairie dog colony. Plague bacteria were most likely introduced to the United States on rat-infested ships from Asia.[142] The first known case of plague was found in San Francisco's Chinatown in 1900. From there, it spread through the West like ripples in a pond when you throw a rock into water.

Figure 9.5. Prairie dog
Photo: Dean Biggins

Prairie dogs have now declined by 97–98 percent of original population numbers throughout North America, as documented by the USFWS and US Geological Survey (USGS).[143] That numerical decline, combined with extreme susceptibility to plague and a lack of regulatory mechanisms by managing agencies, should grant prairie dogs federal protection. Any one of those three conditions legally can qualify a species for protection under the Endangered Species Act. However, because prairie dogs are a politically controversial species, the USFWS has largely avoided protection for three of the four species of prairie dog in the United States. Only Utah prairie dogs (*Cynomys parvidens*) are classified as endangered. As mentioned earlier, keystone species push a system to complexity in nature's economy, whereas for profit in the human economy, people must simplify a system for control.[144] One bison cow–calf unit eats as much vegetation as 288 prairie dogs, so the benefit is not worth the cost of poisoning.[145] Poisoning, however, is often subsidized with taxpayer money.

A grassland inhabited by prairie dogs (like bison, aggregate grazers) is richer and provides a greater mosaic of vegetation structure, an abundance of prey for predators, burrow systems, and altered ecological processes—increased nitrogen content, succulence, productivity of plants, and macroporosity—than grasslands without prairie dogs. In addition, and countering juniper that invade grasslands, prairie dogs clip woody plants, thus limiting conversion of grassland to savanna or shrubland.[146] Studies in Texas showed that elimination of prairie dogs allowed honey mesquite (*Prosopis glandulosa*) to expand at a rate of nearly 15 percent each year.[147]

Prairie dogs thus fit the general classification of a highly interactive species.[148] They affect ecosystem structure, function, and composition in a way that is not wholly duplicated by any other species. The keystone concept means prairie dogs must be protected for more than their own intrinsic value. While intrinsic value is important, so is the impact on other species and processes. The true measure of recovery should be more than taxonomic presence. It should be to restore numbers and geographic distribution of highly interactive species to a point where they can exert their ecological function.[149] A population can be taxonomically present, but without sufficient numbers and a wide enough geographic distribution, it can still be functionally extinct. Thus far, both prairie dogs and bison are too limited in numbers and distribution to maintain their functional role in grasslands. When ecosystems lack functional species and processes, they degrade—as we see on many western grasslands.

Plants

Silverleaf sunflower (*Helianthus argophyllus*), prairie sunflower (*Helianthus petiolaris*), Greek valerian also called Jacob's ladder (*Polemonium caeruleum*), corn daisy (*Glebionis segetum*), trailing fleabane (*Erigeron flagellaris*), and gayfeather (*Liatris spicata*) bloom in early August. Curly cup gumweed (*Grindelia squarrosa*) now has both its yellow flower and gum in cups. The resin of gumweed has a long history of medicinal uses, including for bronchitis, asthma, and as an expectorant. Bees are important pollinators of curly cup gumweed, but the plant does not result in tasty honey. Horses grazing in a field with a lot of gumweed will have resin dots along their noses. Yellow sunflowers and purple gayfeather are pretty much the last blooms on the grasslands at this time of the year.

August 11 to August 20

Birds

In mid-August, lark sparrows, flycatchers, warblers, and lark buntings leave us, and northern birds start to arrive as they pass through. Waterfowl begin to head south for the winter. Fifteen years ago, we could see a hundred mourning doves (*Zenaida macroura*) sitting on the fence at the highway as they flew south. Now there may be only several dozen. Mourning doves are a popular game bird for hunting, but Seamans, Rau, and Sanders showed a gradual decline in the Central and Western United States from 1966 to 2013 in their report *Mourning Dove Population Status*.[150] Meyer found that nest success and nest predation had changed little in his Utah study site from 1952 to 1994.[151] He did find, however, that nest density declined by 20 percent and reproductive output declined by 12–19 percent. The number of fledged young per breeding pair declined by

36 percent.[152] This is a typical result for stressed populations. Mourning doves breed on the refuge. Perhaps the decline we see in August when doves fly south is the result of something in the northern birds.

Figure 9.6. Mourning dove
Photo: Dean Biggins

Mammals

In mid-August, chipmunks actively gather food for winter. Gunnison's prairie dogs and ground squirrels prepare for hibernation. Adult males first enter hibernation in August, followed by adult females about three weeks later. Juvenile males and females stay active until late October. Thinner animals go underground later than well-fed animals. Thirteen-lined ground squirrels are intolerant and aggressive as hibernation approaches.

Plants

Summer is almost over. The warm season grass now puts most plant energy into seeds. The refuge will see some more leaf growth, but less of it now that energy is going into seed production. If the rainy season comes too late, there is less forage for wildlife. Sleepy grass (*Achnatherum robustum*) has now formed seed heads. Sleepy grass contains microbes in the form of fungi and bacteria. The fungi make alkaloids that help the plant during drought, but the alkaloids can be toxic to grazers in high doses. The effect of sleepy grass on many animals is drowsiness, hence the name. In times past, some people fed small amounts of sleepy grass to cattle during a cattle drive. That made the animals easier to control. Sand dropseed grass (*Sporobolus cryptandrus*) also seeds in mid-August. Sand dropseed is an important binder in sandy soils, particularly in the Southwest. Roots can extend about 2 feet (0.6 meters) laterally and to a depth of 8 feet (2.4 meters). It is a perennial warm season grass and is very tolerant of drought. Canadian wild rye (*Elymus canadensis*) and deer muhly (*Muhlenbergia rigens*) form seed heads. Both species are perennial native bunchgrasses.

August 21 to August 31

Plants

Chokecherries, thimbleberries, and Rocky Mountain juniper berries ripen in late August. Thimbleberries are a favorite food for wildlife, and humans like them too. They taste like a red raspberry except they are hollow, like a thimble. For that reason, they fall apart more quickly than a raspberry. The leaf looks like a maple leaf, the plant has no thorns, and the flower is white and pretty. Chokecherries have a reputation for being bitter, but they make a nice pie or jam, particularly if mixed with blueberries. Both chokecherries and thimbleberries can be a boom one year and a bust the next.

Sunflowers are common in late August, and they are aptly named. They have a bright yellow flower, and they track the sun as the Earth turns, facing east in the morning and west by afternoon. I once found a nice patch of sunflowers with a dozen broad-tailed hummingbirds poking their beaks quickly into the center of the flowers. They were catching small insects. Hummingbirds will also snatch insects out of spider webs.

Little bluestem (*Schizachyrium scoparium*) turns red for the fall. The name comes because the stems are a bluish color earlier in the year. Little bluestem is more common in a midgrass prairie, so a dry year in the Southwest produces limited amounts. We mostly see them at the west end of the refuge. During wet years, we can also see a few big bluestem (*Andropogon gerardii*) plants. Big bluestem is also called turkey foot because the seed head resembles the foot of that bird. It is also a warm season grass but is most common in tallgrass prairie. Snakeweed in full flower means that it's time for elk and deer to shed velvet from their antlers. You should listen for early bugling, particularly at sundown after the first of September. The first yellow leaves appear in very small patches of some cottonwood trees.

Figure 9.7. Oak leaf
Artwork: Mary Miller

Insects

Arizona sister butterflies (*Adelpha bredowii*) fly in the oak. You will most commonly see them in Falcon Canyon, where there is water and abundant oak to supply them with food. Arizona sisters are part of the family Nymphalidae, so they are related to monarch and queen butterflies. Checkered skippers (*Pyrgus* spp.) are common on the refuge as well. I leave it classified to genus only because white checkered skippers and common checkered skippers are impossible to separate in the field. They fly low to the ground and seem to like blue flax. Orange and black soldier beetles (family: Cantharidae) are commonly seen in late August on fall wildflowers, particularly curly cup gumweed and other yellow flowers. Soldier beetles are pollinators, and they also feed on aphids.

CHAPTER 10

September

September 1 to September 10

Birds

Many of our summer residents fly south in September. Nesting warblers, yellow breasted chats, and lesser goldfinches are scarce. Rufous hummingbirds have left our feeders and started their southward migration. This is the time that Wilson's warblers (*Cardellina pusilla*) and Townsend's warblers (*Setophaga townsendi*) begin arriving from the north in significant numbers as they pass through to winter range. The Wilson's warblers are easy to see because they flit around in the understory. Their yellow body and the males' black cap make them stunning birds. Wilson's warblers have declined by 60 percent in the last fifty years.[153]

There are still grassland sparrows in September, but they move around in large groups. It can sometimes be difficult to tell when a species leaves if others of the same species replace them during migration from the north. Indeed, some species may move into an area for the winter after summer residents of that same species have moved south. The impression can be that the species is a year-round resident. Marking individuals is a sure way to know.

The prairie falcons and peregrine falcons have departed the refuge by early September. American kestrels are still commonly seen and are found year-round here, although there are reports that they are beginning to decline in the eastern United States. Kestrels are the smallest falcon. Unlike other falcons, kestrels can hover in the air, which increases their hunting efficiency of small birds and small mammals.

Figure 10.1. American kestrel
Photo: Luis Ramirez

Mammals

During this time in September, male antlers are solid and have no velvet. Elk bugling advertises their maleness. When an elk bugles, he lifts his muzzle, stretches his neck, and with great force begins his song. It starts with a note that is more or less in the middle of the sound range but then jumps to a high-pitched squeal that lasts five seconds or so. The squeal is followed by a half dozen low pitched grunting yelps. You can hear a bugle when an elk is a mile (1.6 kilometers) away. Hearing the sound is a real thrill. It signals the start of rut, and it is meant to claim a harem and keep other males away. Males also spray urine on their belly to advertise, and they will wallow in mud. If you come across a wallow, you can smell the elk. The elk rut is at the same time as the pronghorn rut.

Figure 10.2. Elk bugling
Photo: Dean Biggins

A dominant male elk or pronghorn may have ten to thirty females in the harem, but some harems may only have two or three females. The peak of the rut for elk and pronghorn is late September into early October. Hunting season corresponds to this time. In my opinion, shooting a dominant male should not be permitted until after the rut ends around mid-October. For elk, this would mean not shooting a six-point bull. Shooting a dominant male during the rut disrupts breeding. Males of both species try to keep females in a harem, but females do leave and join other harems. By now, the fawns are weaned. Because of the investment females make in carrying the offspring and the even heavier cost of lactation, they choose a mate carefully.

Male pronghorn and elk use a lot of energy trying to keep females in the harem as well as chasing other males away. The pheromones of a female in heat can attract males from over a mile (1.6 kilometers) away. Thus, there will be other males hanging around the harem. Keeping them at bay, while also tending females in heat, leaves little time for eating, and males at the end of a rut can be very depleted. If the rut is followed by a harsh winter, the breeding males may not survive. We have an easily visible male pronghorn at the refuge near Highway 161, and his harem usually contains nine or ten females.

September 11 to September 20

Birds

In mid-September, starlings form large groups, or murmurations, and the tightly packed flock curves and swoops as if they were choreographed. The movement of the flock resembles the movement of a tight school of tropical fish. The flight is often near sunset, and starlings roost in large groups for the night. Warblers and other neotropical migrants move from the north toward their winter homes.

Figure 10.3. Swainson's hawks
Photo: Dean Biggins

Swainson's hawks are around as they head for their winter areas in Argentina. Many Swainson's hawks spend (our) summer on the high plains of the West, then fly to Patagonia to spend (our) winter. One can sometimes see large groups of Swainson's hawks, called kettles, migrating. These groups may include turkey vultures, but the turkey vultures leave the group in Mexico to stay there for the winter. The Swainson's hawks continue to Patagonia in a journey that can be up to 6,000 miles (9,656 kilometers) one-way and take two months. When nesting in the United States, Swainson's hawks eat rodents and rabbits, but in Patagonia, their wintering diet is grasshoppers. In the early 1990s, somewhere between twenty thousand and thirty-five thousand Swainson's hawks died in a single season from pesticide poisoning.[154] The pesticide, monocrotophos, was banned in the United States but was still used widely in Argentina, and the hawks were eating poisoned grasshoppers. In 1996, Argentina banned the pesticide, and Swainson's hawk numbers have since stabilized.

Occasionally, we can have an early cold streak at the refuge, but on September 9, 2019, we had a severe weather event. The temperature went from a high of 91 degrees Fahrenheit (33 degrees Celsius) on September 7 to a high of 36 degrees Fahrenheit (2 degrees Celsius) two days later. The low temperature overnight was 29 degrees Fahrenheit (-1.67 degrees Celsius). Hummingbirds, however, have a strategy to survive a cold streak. They enter torpor overnight, slowing their metabolism and dropping their body temperature to 38 degrees Fahrenheit (3 degrees Celsius). That is a lower temperature than any other bird or nonhibernating mammal can achieve. At daybreak when they can start to feed, they increase their metabolism and raise their body temperature to 104 degrees Fahrenheit (40 degrees Celsius). Their small size would be fatal if they needed to maintain a high body temperature all night, so this extreme form of torpor helps them survive the night-time cold snaps. Sure enough, as soon as the sun rose with a temperature of 29 degrees Fahrenheit (-1.67 degrees Celsius) at daybreak of September 9, the hummingbirds were congregating at our feeders to drink sugar water.

Other migrating birds were not as fortunate as the hummingbirds. The snow stressed them at a time when migration takes a great toll on energy. This caused the largest fall die-off of small migrating birds in New Mexico history. Ornithologists at University of New Mexico and New Mexico State University reported that dead birds were little more than feather and bone. One ornithologist said she saw a bird die while flying and watched it fall to the Earth. Two years later, bird numbers around here are still low.

We have been monitoring grasshoppers at Rio Mora NWR to see the effects of climate change. In June 2019, we counted 1,058 individuals on the grassland transects. A week after the 2019 snow, we counted eleven individuals on the same transects. Birds that survived the stress of the snow had no food for recovery and starved. The grasshopper results we saw here on Rio Mora NWR were replicated throughout the state.

Mammals

Male elk have swollen necks and begin to spar. They become very vocal, and their bugling is heard frequently near sundown. Beavers begin to prepare for winter by storing cottonwood and willow branches in caches. This is also a time when beavers make new dams and thus cut down more trees. Black bears start to build up body condition for hibernation. They can eat up to twenty thousand calories a day. If the acorn and berry crop is poor, it will not be a good sign for bears next spring. If fat stores are insufficient, females will absorb the blastocyst that is dormant in delayed implantation and will not become pregnant. The layers of fat bears put on will be their food for the winter months.

Plants

Virginia creeper (*Parthenocissus quinquefolia*) and poison ivy (*Toxicodendron radicans*) leaves turn bright red in mid-September. Poison ivy causes a rash for humans on contact. The saying for humans is "leaves of three, let it be." Wildlife species don't have an allergic reaction to the oil of the plant. Some even eat the leaves, and birds are attracted to the white berries.

Acorns are fully sized and turn from green to brown. Cockleburs (*Xanthium strumarium*) have their burs developed, and they get almost hopelessly entangled in mammal hair. The burs are still green. The plant is toxic to mammals. The cocklebur is native to North America, but it is invasive. Sagebrush pollen is in the air and stirs allergies for many people.

So does ragweed (*Ambrosia artemisiifolia*) pollen, which can last from mid-August to mid-October. The days rapidly become shorter, and that slows the production of chlorophyll. Thus, yellow or red pigments become more visible as green chlorophyll is reduced. Rose hips are plentiful at this time of year on the refuge. You can add about twenty-five rose hips to a pint of water to make a tea that is high in vitamin C.

Insects

As we noted earlier, fall adult grasshopper species lay eggs this time of year, and those eggs overwinter underground. According to David Lightfoot, this is the strategy for most grasshopper species. Fall grasshoppers were declining here over the few years before 2019, probably because of drought and disrupted rain events. The fall grasshoppers rely on reliable rain patterns for reproduction, and climate change has altered the timing and scale of rain events. The early snow, however, caused an unprecedented fall grasshopper die-off. This dearth of fall grasshoppers extended into 2021, as there were very few fall grasshoppers left to lay eggs, which they should have been doing in mid-September.

We usually see the mid-stage instars of the spring adult grasshopper species in September and October, but after the snow they were also scarce. The 2021 grasshopper counts showed about a fifth of what we saw in June, and there was also a greatly reduced grasshopper biomass. The large spring adult grasshoppers were not around, so their instars were also affected by the fall 2019 snow. Most of the early grasshoppers were small. The long-billed curlews rely on ample grasshopper numbers for reproduction. Whereas in earlier years, watching a curlew walk and feed, we could see a curlew grabbing a grasshopper every few steps. While watching the nesting pairs of curlews in 2021, it looked like they were grabbing a grasshopper every 10 or 15 feet (3 or 5 meters) or so. Of the three long-billed curlew nests that I monitored in 2021, two failed.

September 21 to September 30

Fall Equinox

This is the time of the fall equinox, when the day and night each last twelve hours. After that date, the nights become increasingly longer than the days. The fall equinox marks the end of summer and the start of autumn. The sun passes over the equator on its journey toward shining directly on 23.5 degrees latitude south. That day (winter solstice) will be the depth of winter here, and the depth of summer south of the equator. So, after the fall equinox, nights get cool, but day temperatures can still be hot. Dress in layers to be prepared.

Birds

Rodents aren't free from predators when snakes go into hibernation. In addition to wintering hawks and weasels, greater roadrunners (*Geococcyx californianus*) rely on rodents and small birds for winter prey. Many think that the roadrunner specializes in lizards, but they are generalist enough in diet to survive the winters here. Roadrunners live most of their life on the ground, but they can fly short distances. Their strong beak and ability to run up to 20 mph (32.1 kph) makes them efficient predators. The male and female share incubation duties, and their charisma makes them the New Mexico state bird.

Plants

Three-leaf sumac leaves turn yellow. Milkweed dries and turns brown as well. When dead, milkweed disappears above ground, but the roots remain viable, and the plant can sprout from them the next spring. Cattails (*Typha latifolia*) are also browning, and their cottony seed is visible. Cattails have edible parts—the lower parts of leaves and stems. Flowers rapidly diminish on Rio Mora NWR in late September. There are sunflowers and butterweed (*Senecio spartioides*) still in bloom, and they are yellow—depending on rain. Showy daisies (*Erigeron speciosus*) give us some purple color.

Piñon pine nuts are ready to eat, and they provide nutritious food.

If conditions are not good, the nut will dry up in the shell and be useless. Wildlife are not the only ones to feed on these nuts. They have been an important food source for humans in the Southwest for more than a thousand years. As mentioned earlier, a pound (0.5 kilograms) of these nuts can provide three thousand calories. Virginia creeper berries are also on the vine, but be careful. The berries are toxic for humans and can kill you. The sap can irritate your skin. Birds and mammals, however, can eat the Virginia creeper and juniper berries. Humans don't eat juniper berries, but we do use them to make gin.

Insects

The red-winged grasshopper (*Arphia pseudonietana*) is commonly seen at the refuge in the fall. This grasshopper makes a rapid clacking sound in flight that can resemble the rattling of a rattlesnake. Many people, including biologists, have been fooled by the sound of red-winged grasshoppers flying by.

Reptiles

Snakes have been eating mice and lizards all summer and perhaps bird eggs earlier in the summer. They have been on their summer territory, but it is now time for them to make a trip to their winter den. Some of these trips can be several miles. When coming out of hibernation, rattlesnakes orient themselves and head in a straight line for their summer area. In fall, they return by the same route. People have seen rattlesnakes swimming across a reservoir at this time of year. Rattlesnakes are very loyal to a den site. The path for the ancestor rattlesnakes may have been over ground that was flooded by the dam. The offspring have kept the same route over the years, even if it involves swimming.

These dens can be in rock crevices or underground in burrows. Sometimes, there are multiple snakes communally forming a ball in the den. Other times, there is a single snake in a winter den. When working on black-footed ferret recovery, we were once on a prairie dog colony that served as a rattlesnake hibernaculum. Every time you walked past a burrow opening, you could hear a rattle from inside the tunnel.

The most common rattlesnake on the refuge is the prairie rattlesnake (*Crotalis viridis*). That species is not particularly big and not particularly aggressive. A friend looked at defensive behavior in prairie rattlesnakes. He found that the main defense is to flee (about 50 percent of the time) or lie cryptic (about 25 percent of the time). About 25 percent of the time, they shook their rattle as a warning. They struck at him about 1 percent of the time, but he cautioned that smaller snakes and snakes close to the den would strike more often than adult snakes away from the den. When humans are bitten by rattlesnakes, it frequently happens to young, male humans, especially when alcohol is involved—*hold my beer and watch this.*

CHAPTER 11

October

October 1 to October 15

Birds

Sage thrashers (*Oreoscoptes montanus*), yellow-headed blackbirds (*Xanthocephalus xanthocephalus*), white pelicans (*Pelecanus erythrorhynchos*), snow geese, Canada geese, sandhill cranes, ducks, and other groups of birds fly over on their way south for the winter. Very close to October 1, the turkey vultures that nest and roost at the refuge leave for Mexico, and we no longer see hummingbirds at the feeders.

Northern harriers (*Circus cyaneus*) arrive for the winter in early October. They hunt by flying close to the ground with wings held in a slight V while tipping from side to side. Females are brown, and males are a gray with chocolate wing tips. Both sexes have a white patch at the base of their tail. Harriers mostly eat small mammals, as their talons and grip are not as robust as other raptors. I once saw a female northern harrier land on the back of a duck to try and pull it out of the water. She failed, and after about thirty seconds she let go of the duck and flew away.

Figure 11.1. Northern harrier
Photo: Luis Ramirez

Figure 11.2. Female black bear making a den
Photo: Automatic camera set by Chris Wemmer

Mammals

October is the peak of the rut for elk and pronghorn. It will soon start to taper and will be finished by mid-October. Black bears prepare their dens by pulling in vegetation and twigs to serve as a mat on the floor. We have set automatic cameras on dens to watch the process. The young may join the mother to help drag vegetation into the den. Bear dens are often tight at the entrance. It can be hard for a human to squeeze through, but somehow black bears do it. They often choose rock cavities for dens, but they can also use other crevices, including hollow logs.

"Got your elk?" or "Got your wood?" is a common greeting this time of year. This is the time of year for elk hunting. It's also when people cut wood to burn in a wood stove over winter. Wood is an important source of heat in northern New Mexico. Aldo Leopold once said that wood warms you three times: Once when cutting it, once when splitting it, and once when burning it. We burned wood in the Rio Mora NWR office building.

Insects and Arachnids

The most common butterflies now are the sulphurs, but painted ladies, black swallowtails, red admirals, and monarchs also fly through from the north. Grasshoppers are declining in number but can still be active, particularly the nymphs of the adult spring grasshopper species. Those nymphs will spend the winter under litter.

In October, male New Mexico tarantulas (*Aphonopelma* spp.) move about to search for mates. (Tarantulas are not an insect; they are arachnids with eight legs instead of the six on insects.) Many think this behavior is a migration, but it's not. It's courtship. Females emit pheromones, and then the males seek out their burrows during the night. If a male finds a female lair and courtship is successful, she can lay up to a thousand eggs. If she isn't interested in the male, she may kill and eat him. These movements make mating a risk for the males, so they may only live a couple of years. Females, however, stay close to their burrows, and with less risk, they can live ten to twenty years. Despite their fierce appearance, tarantula venom is harmless to humans. They may bite if harassed, but the bite only feels like a bee sting. If you see tarantulas out at night, just leave them alone.[155] They are docile. Some people like to keep them as pets, even though they aren't cute and cuddly.

Fish

Brown trout (*Salmo trutta*) start to spawn in early October. Brown trout were first introduced to the United States in 1883 for sport fishing. They can tolerate warmer waters and lower oxygen levels than can other species of trout. The native trout in this part of New Mexico is the Rio Grande cutthroat trout (*Oncorhynchus clarkia virginalis*), but that species is limited to about 10 percent of its original range, mostly in headwaters. Native cutthroat trout have declined because of habitat changes, water infrastructure, hybridization with introduced rainbow trout (*Oncorhynchus mykiss*), and competition with brown trout.

October 16 to October 31

The flowers are mostly dry now. All migrating songbirds from the north have moved south for the winter. Sharp-shinned hawks and Cooper's hawks left with the small birds, which are their prey. Goshawks, however, can come to the refuge during winter as they move down from higher elevations. There are still juvenile rock squirrels above ground, but many will disappear soon. That said, some stay active during winter days. Harvester ants have gone below ground for the winter as well.

Fall colors are fading from the oak leaves. They change from green to orange to brown. The leaves of Gambel's oak are brown and dropping, but the leaves of scrub or wavyleaf oak stay attached all winter. Cottonwood leaves are yellow, as the chlorophyll has been reduced to the point where green is no longer visible. There are still waterfowl coming to ponds in the area.

The rut for elk and pronghorn is over. Male elk return to bachelor groups. One day around this time in October, we saw sixty-five pronghorns in a group. There were six dominant males. That would mean there were roughly an average of ten females in the harem of each male, but males are no longer competing with each other, and all is friendly.

Temperatures are falling, but you can still see flies, grasshoppers, and wasps, at least during midday. Even though temperatures are lower, the ground still holds some heat. The bitter cold won't come until that ground heat has dissipated totally, usually in late December and January.

CHAPTER 12

November and December

November 1 to December 31

Gathering piñon nuts is a big event in New Mexico, as they are a nutritious food, but by this time of year, piñon picking is largely finished. Personally, fall is my least favorite season. The warmth of summer is leaving, flowers have died, and the bird songs have gone south. At the beginning of November, there are still a few red-shanked grasshopper nymphs, but the refuge is very still. By the end of the month, the leaves of deciduous trees are dead and have fallen to the ground. Milkweed has disappeared above ground, as they wait for the roots to resprout the plant next spring.

Birds

December marks the winter solstice. It is the shortest day of the year, and thus the longest night. Small birds need to expend a lot of energy to survive that long night, and the short day makes it difficult. They need reliable access to high quality food, which they sequester in caches.

Migrating bald eagles arrive in early December. A dozen or so stay by a pond near Buena Vista, New Mexico, and that pond holds ducks all winter. The pond is about 10 miles (16 kilometers) west of Rio Mora NWR. The bald eagles perch in the bare cottonwood trees and probably feed on the ducks and fish of the pond. When the Pueblo of Pojoaque harvest a few bison in December and January, the bald eagles come to the refuge to feed on the gut piles.

Mammals

Black bears hibernate in mid-November, and most dens show evidence of prior use. With good nutrition, the blastocyst of pregnant bears implants at this time, and the fetus develops to be born in January while the female sleeps. The newborn is about the size of a rat and crawls to a teat where it stays until spring. Black bears will spend about 150 days hibernating in a den, during which time they won't eat, drink, defecate, or urinate. Hibernating bears lose about 25 percent of their body weight, which is why the fall crop of mast is important. Their metabolic and heart rates will drop by 50 percent, but body temperature falls only 12 degrees Fahrenheit (6.67 degrees Celsius). That means black bears can come out of hibernation swiftly and be active if their den is disturbed. They can go from hibernation to a dead run very quickly.

Mule deer begin their rut in late November, which is triggered by photoperiod changes. Rut is not triggered by colder weather as some people assume. The males get thick necks and spar with other males over access to females. During the rut, males are driven by hormones and are less cautious. They are thus easier to kill, although once again it doesn't seem right to disrupt mating by killing dominant males. Females will stay in estrus for about three days. If they do not breed, the estrus cycle will return in about three weeks. Winter diet of mule deer is about 75 percent browse. When grasses and forbs emerge in spring, the amount of browsing drops to about 50 percent. Earlier, I mentioned that coprophagia is a way for rabbits, picas, and beavers to eat their own feces and thus extract more nutrients. Deer and elk increase their ability to absorb more nutrients by rumination. Regurgitating the cud to chew it a second time further breaks down plant matter and aids digestion.

Winter snows offer a naturalist the chance to follow tracks easily. Interpreting tracks can tell you the size of the animal, the pace used while moving, and whether the animal was heading directly toward its goal or searching for prey in a zig-zag motion. If you follow tracks, you will see how hard a carnivore must work to feed itself. Coyotes need about 1.25 pounds (0.57 kilograms) of meat a day during winter. Coyotes and foxes both dine heavily on rodents and rabbits. Both have excellent hearing to locate voles in subnivean tunnels. When snow is deep, both species leap into the air and dive headfirst into the snow to grab the prey.

Canid tracks, like coyote or fox, are different than cat tracks. While canid tracks may show claw marks, that is not always the case. A canid track will show an X pattern. You will be able to draw a line along the outside of a pad, and that line will extend between the two toes on the inside of the track. You can then draw a line along the inside of the pad, which will extend between the two toes on the outside of the track. That will show the X. Feline tracks have a wider pad and track, so it is impossible to draw the same X as you can with canids. In Figure 12.1, the top track is canid and the bottom is felid. A wolf track will have the top of the outer toes at or slightly below the bottom of the inner toes, whereas a dog will have the top of the outer toes above the bottom of the inner toes.

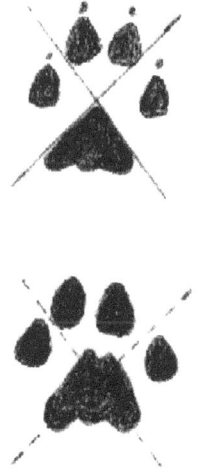

Figure 12.1. Comparison of canid (*top*) and felid (*bottom*) tracks
Artwork: Brian Miller

Long-tailed weasels (*Mustela frenata*) have a high surface area to body mass ratio because they are long and slender. That high surface area to body mass ratio means that they easily lose heat when on the surface in winter. Thus, long-tailed weasels must eat about 30 percent of their body weight each day to maintain themselves. Their diet is largely rodents. The benefit of being long and thin is that weasels can enter tunnels and hunt in places that would be difficult for other species.

Bobcats (*Lynx rufus*) need about a pound (0.5 kilograms) of meat per day during winter. They rely heavily on rabbits but are opportunistic. In very deep snow, they can even kill a deer. Pumas (*Puma concolor*) need to kill a deer every two weeks to make it through winter. Elk can be a low portion of diets in some areas but make up to 25 percent of puma diet in others.

Figure 12.2. Puma head
Artwork: Brian Miller

I had the good fortune to find a set of tracks showing a puma killing a deer at Rio Mora NWR. The ground revealed a mixed set of running deer tracks followed by puma tracks. Suddenly the puma tracks were no longer seen. At that point, the deer tracks began sliding outward from the weight of the puma on the deer's back. Then the puma tracks reappeared, and the deer tracks stopped sliding and resumed running. About 30 feet (9 meters) later, the puma tracks disappeared, and the deer tracks started sliding again. After a few sliding steps, the deer was on the ground. The drag marks made by the puma pulling the deer led to a brushy patch, and indeed the puma had stashed the carcass under those bushes. We also occasionally see a place where pumas repeatedly place carcasses. One such place had bones from three elk. In Mexico, such places are called *comedores*—dining rooms.

In cold areas, badgers can stay underground for more than seventy consecutive days. Hank Harlow found that badgers in Wyoming enter torpor while underground. One individual entered torpor on thirty occasions while below ground. Each of these cycles lasted about thirty hours. During these bouts of hibernation, the animal's heart rate slowed by 50 percent and its body temperature dropped about 29 degrees Fahrenheit (16 degrees Celsius). It is most likely the same for northern New Mexico. You will see badgers much less often in winter.[156]

We discussed the role of mutualists—piñon jays and insect pollinators—and ecosystem engineers—beaver, bison, and prairie dogs—as highly interactive species. Predators are highly interactive species as well. Carnivores and other types of predators sit at the top of the food chain. From that position, they are important drivers of ecosystems. In a nutshell, herbivores can reduce the biomass of plants, but in turn, the herbivore biomass is held in check by carnivores. In 75 percent of roughly three hundred studies, systems that evolved with predators declined when they were removed. Without predators, herbivore numbers increased, and that increased plant damage, decreased plant biomass, and decreased plant-reproductive output. Without top predators to check the numbers of mesopredators, the mesopredator release changes the structure of smaller mammals and birds, reducing diversity in those communities.[157]

Without wolves, Yellowstone elk became lazy and spent their time near the streams. This lack of movement by elk took a heavy toll on aspen and willows near streams, which are important nesting habitats for migratory birds. Joel Berger reported that there are higher numbers of nesting songbirds in areas with wolves and grizzlies than in areas without because the top predators checked the herbivore numbers and forced herbivores to move more widely away from streams. Without wolves, beavers had no willows and aspen for food, shelter, and dams. Without beavers, the river began to incise and lost its ecological function. When wolves returned to Yellowstone, they reversed these negative effects. The world is green because of predators.[158]

Figure 12.3. Wolf
Artwork: Brian Miller

Concluding Thoughts

"It's not that I have just seen the mountains, I have felt them. First came the road. With the road came settlements. With settlements came civilization. With civilization, the jaguars went away."

—Humberto Rangel to me in 1995 on the Chamela-Cuixmala Biosphere Reserve

Figure 13.1. Humberto Rangel
Photo: Brian Miller

In 1995, we started a project in Jalisco, Mexico, to monitor jaguars (*Panthera onca*) and pumas so that we could know the conservation needs of both around the Chamela-Cuixmala Biosphere Reserve. That project continues today under the guidance of Rodrigo Nuñez, who used it to earn his PhD. By the way, in the 1930s a jaguar was shot and killed about 50 miles (80 kilometers) north of the Rio Mora NWR near Springer, New Mexico. Imagine the thrill of seeing a jaguar in northern New Mexico! If you do, please don't shoot it. Jaguars once roamed the southwestern United States as the northern part of their recent range. During the Pleistocene, that range extended to Oregon. Today, only a few wander north of the United States and Mexico border.

Figure 13.2. Jaguar

Artwork: Brian Miller

Humberto lived in the mountains most of his life. In his youth, he trapped jaguars to sell the pelts. He lived in a *choza*, which is a room built from posts with sticks woven around the posts. The roof was thatch, and he drank water from a nearby spring. When the biosphere reserve started, he turned 180 degrees and became a guard in the mountains to protect wildlife. The refuge gave him a cabin which was comfortable, had a well, but had no electricity. I stopped by his cabin every day while going into the hills. He was a constant source of knowledge about nature. Humberto showed us how to set snares so that we could place radio-collars on the cats. I kept in touch with Humberto after returning to the United States. In 2000, he became too old to live by himself in isolation. The people at the biosphere reserve moved him into the nearby town of Zapata. He died a month later. He wasn't meant for a town.

Be like Humberto. Take some time to admire nature. Don't just look from the car window. Smell the flowers and the recently used elk bed. Be curious. Feel the wind. Feel the cold. Feel the wild. Then protect it. Don't expect the government, or institutions, to do the right thing. *Make* them do the right thing. An active citizenry is essential to change.

Acknowledgments

This writing was inspired by Frank Craighead's book *For Everything There Is a Season*, which told about the unfolding of nature in the Grand Teton National Park.[159]

    I have been lucky to have some people who inspired me while following conservation as a career. Ulie Seal helped me during my PhD when I was working with black-footed ferrets. At one point, there were only ten individual black-footed ferrets left to represent the entire species. Ulie was a strong advocate for captive breeding, which saved the species. Chris Wemmer directed my Smithsonian fellowship to increase the survival of black-footed ferrets born in captivity and scheduled to be released in the wild. He has been a friend ever since. Michael Soulé became a close friend and mentor from 1990 to 2020, when he died of a brain bleed. Michael started the field of conservation biology and was a brilliant mind. I still miss his friendship and the many discussions we had. Dave Foreman was an influential figure in conservation and an unstoppable advocate for nature. He also died too soon, in 2022. The Wildlands Project was the most invigorating and forward-thinking group that I have ever been part of. I miss those interactions. Rich Reading hired me to help him start the conservation and research program at the Denver Zoo in 1997. He is like a brother, and Shantini Ramakrishnan is like a sister.

    I came to the Wind River Ranch in 2005. There I worked with Abran Bolivar Casius (Boli), Theresa Grey, Brooks Read, Sherry Thompson, Eugene Thaw, Katie Flanagan, and Maurice Hornocker. When Wind

River Ranch became Rio Mora National Wildlife Refuge, I worked with Rob Larañaga, Leonard Romero, Luis Ramírez, Shantini, Joe Zebrowski, Erica Garroutte, and the bison team from Pueblo of Pojoaque, including Phil Viarrial, Gabe Montoya, Luke Viarrial, and Dario Caraveo. The team was a dedicated, cooperative, warm, and wonderfully productive group. There are also numerous students, interns, neighbors, and colleagues who contributed to the successes at Rio Mora NWR.

Eugene Thaw donated his land to become part of the US Fish and Wildlife refuge system. When the USFWS had no budget for Rio Mora NWR, Eugene donated $1,715,000 to Denver Zoo to run the programs. When that money expired, Senator Tom Udall provided a budget for Rio Mora NWR from the Senate Appropriations Committee. When the research and education programs at Rio Mora NWR were in danger, New Mexico Highlands University absorbed them into their STEM efforts. Thanks to Sam Minner, Edward Martínez, Kent Reid, Ian Williamson, and Joe Zebrowsky for that.

Dean Biggins, Rich Reading, Chris Wemmer, and John Davis reviewed the manuscript. John Davis, Dave Foreman, and Susan Morgan of the Rewilding Institute encouraged writing this book. Many thanks to Dean, Rich, Chris, Luis Ramirez, Shantini Ramakrishnan, and Mary and Anabella Miller for their photo contributions.

Thanks to Melissa Stevens of Purple Ninja Editorial and Becky Bayne, Becky's Graphic Design®, LLC for their excellent support and advice on editing and layout.

Throughout life I have been blessed with a number of good friends. I will avoid names out of fear that I may overlook someone, but you all know who you are.

Finally, my wife, Carina, and my daughters, Mary and Anabella, have been a constant source of pleasure and joy. Mary and Anabella shared in drawing the figures, which made the book a family effort

APPENDIX A: List of Vertebrate Species

BIRDS

Order: Anseriformes
Family: Anatidae

    Canada goose (*Branta Canadensis*) summer

    Snow goose (*Chen caerulescens*) migrating

    Wood duck (*Alix sponsa*) summer

    Gadwall (*Anas strepera*) migrating

    Mallard (*Anas platyrhynchos*) summer

    American wigeon (*Anas Americana*) summer

    Northern shoveler (*Anas clypeata*) migrating

    Northern pintail (*Anas acuta*) migrating

    Green-winged teal (*Anas crecca*) summer

    Blue-winged teal (*Anas discors*) summer

    Cinnamon teal (*Anas cyanoptera*) migrating

    Canvasback (*Aythya valisineria*) migrating

Redhead (*Aythya americana*) migrating

Ring-necked duck (*Aythya collaris*) migrating

Bufflehead (*Bucephala albeola*) migrating

Common goldeneye (*Bucephala clangula*) migrating

Common merganser (*Mergus merganser*) summer

Ruddy duck (*Oxyura jamaicensis*) migrating

Tundra swan (*Cygnus columbianus*) migrating

Order: Pelecaniformes
Family: Phalacrocoracidae

Double-crested cormorant (*Phalacrocorax carbo*) summer

Family: Ardeidae

Great blue heron (*Ardea Herodias*) summer

Black-crowned night-heron (*Nycticorax nycticorax*) summer

Green heron (*Butorides virescens*) summer

White-faced ibis (*Plegadis chihi*) summer

Order: Galliformes
Family: Phasianidae

Wild turkey (*Meleagris gallopavo*) all year

Order: Accipitriformes
Family: Cathartidae

Turkey vulture (*Cathartes aura*) summer

Family: Pandionidae

Osprey (*Pandion haliaetus*) summer

Family: Accipitridae

Northern harrier (*Circus hudsonius*) summer

Sharp-shinned hawk (*Accipiter striatus*) summer

Cooper's hawk (*Accipiter cooperli*) summer)

Goshawk (*Accipiter gentilis*) summer

Swainson's hawk (*Buteo swainsoni*) summer

Red-tailed hawk (*Buteo jamaicensis*) summer

Ferruginous hawk (*Buteo regalis*) summer

Bald eagle (*Haliaeetus leucocephalus*) winter but nesters stay all year

Golden eagle (*Aquila chrysaetos*) all year

Family: Falconidae

American kestrel (*Falco sparverius*) summer

Merlin (*Falco columbarius*) migrating

Prairie falcon (*Falco mexicanus*) summer

Peregrine falcon (*Falco peregrinus*) summer

Order: Gruiformes

Family: Gruidae

Sandhill crane (*Antigone canadensis*) migrating

Family: Rallidae

American coot (*Fulica americana*) summer

Sora rail (*Porzana carolina*) summer

Order: Charadriiformes

Family: Charadriidae

Mountain plover (*Charadrius montanus*) summer

Killdeer (*Charadrius vociferous*) summer

Family: Scolopacidae

Spotted sandpiper (*Actitis macularia*) summer

Solitary sandpiper (*Tringa solitaria*) migrating

Long-billed curlew (*Numenius americanus*) summer

Family: Laridae

California gull (*Larus californicus*) summer

Forster's tern (*Sterna forsteri*) migrating

Order: Columbiformes
Family: Columbidae

Rock pigeon (*Columba livia*) all year

Eurasian collared dove (*Streptopelia decaocto*) all year

Mourning dove (*Zenaida macroura*) all year

White-winged dove (*Zenaida asiatica*) one sighting

Order: Strigiformes
Family: Tytonidae

Barn owl (*Tyto alba*) all year

Family: Strigidae

Great-horned owl (*Bubo virginianus*) all year

Long-eared owl (*Asio otus*) all year

Short-eared owl (*Asio flammeus*) winter

Western screech owl (*Otus kennicottii*) all year

Saw-whet owl (*Aegolius acadicus*) all year

Order: Cuculiformes
Family: Cuculidae

Greater roadrunner (*Geococcyx californianus*) all year

Order: Caprimulgiformes
Family: Caprimulgidae

Common nighthawk (*Phalaenoptilus nuttallii*) summer

Common poorwill (*Chordeiles minor*) summer

Order: Apodiformes
Family: Apodidae

White-throated swift (*Aeronautes saxatalis*) summer

Family: Trochilidae

Black-chinned hummingbird (*Archilochus alexandri*) summer

Broad-tailed hummingbird (*Selasphorus platycercus*) summer

Rufous hummingbird (*Selasphorus rufus*) migrating

Order: Coraciiformes
Family: Alcedinidae

Belted kingfisher (*Ceryle alcyon*) all year

Order: Piciformes
Family: Picidae

Lewis's woodpecker (*Melanerpes lewis*) summer

Red-headed woodpecker (*Melanerpes erythrocephalus*) summer, rare

Red-naped sapsucker (*Sphyrapicus nuchalis*) summer

Williamson's sapsucker (*Sphyrapicus thyroideus*) summer, rare

Ladder-backed woodpecker (*Picoides scalaris*) all year

Downy woodpecker (*Picoides pubescens*) all year

Hairy woodpecker (*Picoides villosus*) all year

Northern flicker (*Colaptes auratus*) all year

Order: Passeriformes

Family: Tyrannidae

    Ash-throated flycatcher (*Myiarchus tyrannulus*) summer

    Willow flycatcher (*Empidonax traillii*) summer

    Gray flycatcher (*Empidonax wrightii*) summer

    Dusky flycatcher (*Empidonax oberholseri*) summer

    Vermillion flycatcher (*Pyrocephalus rubinus*) rare

    Olive-sided flycatcher (*Contopus pertinax*) migrating

    Western wood-pewee (*Contopus sordidulus*) summer

    Eastern phoebe (*Sayornis phoebe*) migrating

    Black phoebe (*Sayornis nigricans*) summer

    Say's phoebe (*Sayornis saya*) summer

    Cassin's kingbird (*Tyrannus vociferans*) summer

    Western kingbird (*Tyrannus verticalis*) summer

    Eastern kingbird (*Tyrannus tyrannus*) rare

Family: Laniidae

    Loggerhead shrike (*Lanius ludovicianus*) summer

    Northern shrike (*Lanius excubitor*) rare

Family: Vireonidae

    Plumbeous vireo (*Vireo plumbeus*) summer

    Warbling vireo (*Vireo gilvus*) summer

    Cassin's vireo (*Vireo cassinii*) migrating

    Red-eyed vireo (*Vireo olivaceus*) rare

Family: Corvidae

    Clark's nutcracker (*Nucifraga columbiana*) rare

    Steller's jay (*Cyanocitta stelleri*) usually in winter

    Western scrub-jay (*Aphelocoma californica*) all year

Pinyon jay (*Gymnorhinus cyanocephalus*) all year

Black-billed magpie (*Pica hudsonia*) all year

American crow (*Corvus brachyrhynchos*) all year

Common raven (*Corvus corax*) all year

Family: Alaudidae

Horned lark (*Eremophila alpestris*) all year

Family: Hirundinidae

Violet-green swallow (*Tachycineta thalassina*) summer

Tree swallow (*Tachycineta bicolor*) migrating

Northern rough-winged swallow (*Stelgidopteryx serripennis*) summer

Bank swallow (*Riparia riparia*) summer

Cliff swallow (*Petrochelidon pyrrhonota*) summer

Barn swallow (*Hirundo rustica*) summer

Family: Paridae

Black-capped chickadee (*Poecile atricapillus*) all year

Mountain chickadee (*Poecile gambeli*) all year

Juniper titmouse (*Baeolophus ridgwayi*) all year

Red-breasted nuthatch (*Sitta canadensis*) all year

White-breasted nuthatch (*Sitta carolinensis*) all year

Pygmy nuthatch (*Sitta pygmaea*) all year

Family: Aegithalidae

American bushtit (*Psaltriparus minimus*) all year

Family: Troglodytidae

Rock wren (*Salpinctes obsoletus*) summer

Canyon wren (*Catherpes mexicanus*) summer

Bewick's wren (*Thryromanes bewickii*) summer

House wren (*Troglodytes aedon*) summer

Family: Muscicapidae

Ruby-crowned kinglet (*Regulus calendula*) summer

Golden-crowned kinglet (*Regulus satrapa*) rare

Blue-gray gnatcatcher (*Polioptila caerulea*) summer

Family: Cinclidae

American dipper (*Cinclus mexicanus*) all year/rare

Family: Turdidae

Western bluebird (*Sialia mexicana*) all year

Mountain bluebird (*Sialia currucoides*) all year

Townsend's solitaire (*Myadestes townsendi*) all year

Hermit thrush (*Catharus guttatus*) summer

American robin (*Turdus migratorius*) all year

Family: Mimidae

Gray catbird (*Dumetella carolinensis*) summer

Northern mockingbird (*Mimus polyglottos*) summer

Sage thrasher (*Oreoscoptes montanus*) migrating

Brown thrasher (*Toxostoma bendirei*) summer

Family: Motacillidae

American pipit (*Anthus rubescens*) migrating

Family: Sturnidae

European starling (*Sturnus vulgaris*) all year

Family: Bombycillidae

Cedar waxwing (*Bombycilla cedrorum*) winter

Family: Parulidae

Painted redstart (*Myioborus pictus*) rare

Orange-crowned warbler (*Oreothlypis celata*) summer

Virginia's warbler (*Oreothlypis virginiae*) summer

Yellow warbler (*Setophaga petechia*) summer

Grace's warbler (*Setophaga graciae*) summer

Townsend's warbler (*Setophaga townsendi*) summer

Yellow-rumped warbler (*Setophaga coronata*) summer

Cerulean warbler (*Dendroica cerulea*) out of range

Common yellowthroat (*Geothlypis trichas*) summer

Kentucky warbler (*Geothlypis formosa*) summer

Wilson's warbler (*Cardellina pusilla*) migrating

Family: Icteriidae

Yellow-breasted chat (*Icteria virens*) summer

Family: Emberizidae

Green-tailed towhee (*Pipilo chlorurus*) summer

Spotted towhee (*Pipilo maculatus*) all year

Canyon towhee (*Melozone fusca*) all year

Dickcissel (*Spiza americana*) rare

Rufous-crowned sparrow (*Aimophila ruficeps*) all year

Chipping sparrow (*Spizella passerina*) summer

Brewer's sparrow (*Spizella breweri*) migrating

Clay-colored sparrow (*Spizella pallida*) migrating

Vesper sparrow (*Pooecetes gramineus*) summer

Grasshopper sparrow (*Ammodramus savannarum*) summer

Lark sparrow (*Chondestes grammacus*) summer

Song sparrow (*Melospiza melodia*) all year

Lincoln's sparrow (*Melospiza lincolnii*) migrating

Lark bunting (*Calamospiza melanocorys*) migrating

White-crowned sparrow (*Zonotrichia leucophrys*) winter

White-throated sparrow (*Zonotrichia albicollis*) rare

Cassin's sparrow (*Peucaea cassinii*) summer

Dark-eyed junco (*Junco hyemalis*) all year

Family: Cardinalidae

Hepatic tanager (*Piranga flava*) summer

Western tanager (*Piranga ludoviciana*) summer

Black-headed grosbeak (*Pheucticus melanocephalus*) summer

Rose-breasted grosbeak (*Pheucticus ludovicianus*) rare

Blue grosbeak (*Passerina caerulea*) summer

Lazuli bunting (*Passerina amoena*) summer

Family: Icteridae

Red-winged blackbird (*Agelaius phoeniceus*) all year

Western meadowlark (*Sturnella neglecta*) all year

Eastern meadowlark (*Sturnella magna*) summer

Yellow-headed blackbird (*Xanthocephalus xanthocephalus*) migrating

Brewer's blackbird (*Euphagus cyanocephalus*) all year

Common grackle (*Quiscalus quiscula*) summer

Great-tailed grackle (*Quiscalus mexicanus*) all year

Brown-headed cowbird (*Molothrus ater*) summer

Bullock's oriole (*Icterus bullockii*) summer

Family: Fringillidae

House finch (*Haemorhous mexicanus*) all year

Cassin's finch (*Haemorhous cassinii*) winter

Pine siskin (*Spinus pinus*) all year

American goldfinch (*Spinus tristis*) winter

Lesser goldfinch (*Spinus psaltria*) summer

Red crossbill (*Loxia curvirostra*) rare

Evening grosbeak (*Coccothraustes vespertinus*) winter

House sparrow (*Passer domesticus*) all year

Herps and Crustacea

Order: Caudata
Family: Ambystomatidae

Tiger salamander (*Ambystoma tigrinum*)

Order: Anura
Family: Bufonidae

Woodhouse toad (*Bufo woodhousii*)

Red-spotted toad (*Bufo punctatus*)

Family: Pelobatidae

New Mexico spadefoot toad (*Spea multiplicata*)

Plains spadefoot (*Spea bombifrons*)

Family: Hylidae

Canyon tree frog (*Hyla arenicolor*)

Western chorus frog (*Pseudacris triseriata*)

Family: Ranidae

American bullfrog (*Rana catesbeiana*)

Northern leopard frog (*Rana pipiens*)

Order: Testudines
Family: Chelydridae

Snapping turtle (*Chelydra serpentina*)

Order: Squamata
Suborder: Sauria

Family: Crotaphytidae

    Eastern collared lizard (*Crotaphytus collaris*)

Family: Phrynosomatidae

    Pygmy short-horned lizard (*Phrynosoma douglassii*)

    Prairie lizard (*Sceloporus undulatus*)

    Lesser earless lizard (*Holbrookia maculata*)

Family: Teiidae

    Plateau striped whiptail (*Cnemidophorus velox*)

    Six-lined racerunner (*Cnemidophorus sexlineatus*)

Family: Scincidae

    Many-lined skink (*Eumeces multivirgatus*)

    Great Plains skink (*Eumeces obsoletus*)

Suborder: Serpientes

Family: Colubridae

    Coachwhip (*Masticophis flagellum*)

    Eastern racer (*Coluber constrictor*)

    Ringneck snake (*Diadophis punctatus*)

    Corn snake (*Elaphe guttata*)

    Western hognose snake (*Heterodon nasicus*)

    Smooth green snake (*Liochlorophis vernalis*)

    Bullsnake (*Pituophis melanoleucus*)

    Common kingsnake (*Lampropeltis getula*)

    Milk snake (*Lampropeltis triangulum*)

    Blackneck garter snake (*Thamnophis cyrtopsis*)

Western terrestrial garter snake (*Thamnophis elegans vagrans*)

Plains garter snake (*Thamnophis radix*)

Lined snake (*Tropidoclonion lineatum*)

Family: Viperidae

Western rattlesnake (*Crotalus viridis*)

Western diamondback rattlesnake (*Crotalus atrox*)

Fish

Creek chub (*Semotilus atromaculatus*)

Rio Grande chub (*Gila pandora*)

Central stoneroller (*Campostoma anomalum*)

Longnose dace (*Rhinichthys cataractae*)

White sucker (*Catostomus commersonii*)

Brown trout (*Salmo trutta*)

Fathead minnow (*Pimephales promelas*)

Green sunfish (*Lepomis cyanellus*)

Rio Grande cutthroat trout (*Oncorhynchus clarkii virginalis*) has been extirpated

Mammals

Order: Insectivora

Family: Soricidae

Desert shrew (*Notiosorex crawfordi*)

Order: Chiroptera (*Incomplete*)

Family: Vespertilionidae

Little brown myotis (*Myotis lucifugus*)

Hoary bat (*Lasiurus cinereus*)

Western long-eared bat (*Myotis evotis*)

Order: Lagomorpha
Family: Leporidae

Desert cottontail (*Sylvilagus audubonii*)

Black-tailed jackrabbit (*Lepus californicus*)

Order: Carnivora
Family: Ursidae

American black bear (*Ursus americanus*)

Family: Procyonidae

Northern raccoon (*Procyon lotor*)

Ring-tailed cat (*Bassariscus astutus*)

Family: Mustelidae

Long-tailed weasel (*Mustela frenata*)

Striped skunk (*Mephitis mephitis*)

Western spotted skunk (*Spilogale gracillis*)

Common hog-nosed skunk (*Conepatus leuconotus*)

American badger (*Taxidea taxus*)

Family: Canidae

Coyote (*Canis latrans*)

Gray fox (*Urocyon cinereoargenteus*)

Swift fox (*Vulpes velox*)

Family: Felidae

Puma (*Puma concolor*)

Bobcat (*Lynx rufus*)

Order: Rodentia
Family: Sciuridae

  Gunnison's prairie dog (*Cynomys gunnisoni*)

  Thirteen-lined ground squirrel (*Spermophilus tridecemlineatus*)

  Rock squirrel (*Spermophilus variegatus*)

  Colorado chipmunk (*Tamias quadrivittatus*)

  Least chipmunk (*Tamias minimus*)

Family: Geomyidae

  Botta's pocket gopher (*Thomomys bottae*)

Family: Heteromyidae

  Ord kangaroo rat (*Dipodomys ordii*)

Family: Castoridae

  American beaver (*Castor canadensis*)

Family: Muridae

  Hispid pocket mouse (*Chaetodipus hispidus*)

  Plains pocket mouse (*Perognathus flavescens*)

  Silky pocket mouse (*Perognathus flavus*)

  Western harvest mouse (*Reithrodontomys megalotis*)

  Plains harvest mouse (*Reithrodontomys montanus*)

  North American deer mouse (*Peromyscus maniculatus*)

  White-footed mouse (*Peromyscus leucopus*)

  Piñon deer mouse (*Peromyscus truei*)

  Brush mouse (*Peromyscus boylii*)

  Northern rock mouse (*Peromyscus nasutus*)

  Northern grasshopper mouse (*Onychomys leucogaster*)

  White-throated woodrat (*Neotoma albigula*)

  Mexican woodrat (*Neotoma mexicana*)

Common muskrat (*Ondatra zibethicus*)

Family: Phenacomys

Prairie vole (*Microtus ochrogaster*)

Long-tailed vole (*Microtus longicaudus*)

Family: Erethizontidae

North American porcupine (*Erethizon dorsatum*)

Family: Castoridae

American beaver (*Castor canadensis*)

Family: Zapodidae

New Mexico meadow jumping mouse (*Zapus hudsonius luteus*)

Order: Artiodactyla

Family: Cervidae

White-tailed deer (*Odocoileus virginianus*)

Mule deer (*Odocoileus hemionus*)

Elk (*Cervus elaphus*)

Family: Bovidae

Bison (*Bison bison*)

Family: Antilocapridae

Pronghorn (*Antilocapra americana*)

Endnotes

1. Leopold, *Sand County Almanac*, xvii.
2. Cotter, "Eugene V. Thaw, Influential Art Collector."
3. Mac et al., *Status and Trends of the Nation's Biological Resources*, 560.
4. For restoration methods used at the refuge, refer to the Zeedyk publications that are listed in the reference section.
5. Zeedyk, Walton, and Gadzia, *Characterization and Restoration of Slope Wetlands*, 2014.
6. Zeedyk and Jansens, *Introduction to Erosion Control*, 2006.
7. Zeedyk and Jansens, *Introduction to Erosion Control*, 2006.
8. Zeedyk and Van Clothier, *Let the Water Do the Work*, 2009.
9. Zeedyk and Van Clothier, *Let the Water Do the Work*, 2009.
10. New Mexico Department of Game and Fish, *Comprehensive Wildlife Conservation Strategy*, 23.
11. Smith, A, "State of Tribal Co-Management."
12. MacArthur and Wilson, *Theory of Island Biogeography*, 1967.
13. Dick-Peddie, *New Mexico Vegetation*, 1993; US Fish and Wildlife Service, *Rio Mora National Wildlife Refuge*, 8.
14. US Fish and Wildlife Service, *Birds of Conservation Concern*, 2008.
15. Chronic, *Roadside Geology of New Mexico*, xiii; Smith and Siegel, *Windows into the Earth*, 34.
16. Smith and Siegel, *Windows into the Earth*, 34.
17. Chronic, *Roadside Geology of New Mexico*, 9 and 168; Price, *Geology of Northern New Mexico's Parks*, 2010.
18. Chronic, *Roadside Geology of New Mexico*, 165–166.
19. Zeedyk and Jansens, *Introduction to Erosion Control*, 1.

20 New Mexico State University, "Loamy Upland."
21 Hewitt, "Ice Ages: Species Distributions, and Evolution," 339–361.
22 Martin and Klein, eds. *Quaternary Extinctions*, 354-403.
23 Martin, *Twilight of the Mammoths*, 2007; Ward, *Call of the Distant Mammoths*, 1997.
24 Ward, *Call of the Distant Mammoths*, 194–195.
25 Flannery, *Future Eaters*, 195–197, 243–244.
26 Flores, *Wild New World*, 102.
27 Flores, "A Long Love Affair," 11.
28 Ortiz, "Popay's Leadership," 107–113.
29 Brown, *Bury My Heart at Wounded Knee*, 1970.
30 Fletcher and Robbie, "Historic and Current Conditions," 120–129.
31 Fletcher and Robbie, "Historic and Current Conditions," 120–129.
32 Fletcher and Robbie, "Historic and Current Conditions," 120–129.
33 Lott, *American Bison*, 69; Franke, *To Save the Wild Bison*, 53.
34 Lott, *American Bison*, 175–179.
35 Block and Finch, "Songbird Ecology in Southwestern Ponderosa Pine Forests," 45–46.
36 Aby, "Date of Arroyo Cutting," 76.
37 Matthiessen, *Wildlife in America*, 183–191.
38 Terborgh and Estes, *Trophic Cascades*, 2010.
39 Branch et al., "Genetic Basis of Spatial Cognitive Variation," 210–219; Mason, "Total Recall."
40 Mason, "Total Recall."
41 Fitzgerald, Meaney, and Armstrong, *Mammals of Colorado*, 317–320; Pelton, "Black Bear (*Ursus americanus*)," 547–555.
42 Fitzgerald, Meaney, and Armstrong, *Mammals of Colorado*, 317–320; Pelton, "Black Bear (*Ursus americanus*)," 547–555.
43 Tom Beck, personal communication with author.
44 Fitzgerald, Meaney, and Armstrong, *Mammals of Colorado*, 302–305, 310–312, 315–316.
45 Fitzgerald, Meaney, and Armstrong, *Mammals of Colorado*, 225–229; Baker and Hill, "Beaver (*Castor canadensis*)," 288–306.

46 Baker and Hill, "Beaver (*Castor canadensis*)," 288–306.
47 Fitzgerald, Meaney, and Armstrong, *Mammals of Colorado*, 225–229; Baker and Hill, "Beaver (*Castor canadensis*)," 288–306.
48 Fitzgerald, Meaney, and Armstrong, *Mammals of Colorado*, 225–229; Baker and Hill, "Beaver (*Castor canadensis*)," 288–306.
49 Zeedyk and Van Clothier, *Let the Water Do the Work*, 38 and 80.
50 Zeedyk and Van Clothier, *Let the Water Do the Work*, 20–21.
51 Zeedyk and Van Clothier, *Let the Water Do the Work*, 2009; Zeedyk, *Introduction to Induced Meandering*, 1.
52 Foreman, *Rewilding North America*, 65.
53 Sibley, *Sibley Guide to Bird Life and Behavior*, 344.
54 Hawk Mountain Global Raptor Conservation, "Golden Eagle."
55 Hawk Mountain Global Raptor Conservation, "Golden Eagle."
56 Sibley, *Sibley Guide to Bird Life and Behavior*, 15.
57 Sibley, *Sibley Guide to Bird Life and Behavior*, 21–24.
58 Black, "Why Bats Are One;" Speakman, "The Evolution of Flight," 111.
59 Sibley, *Sibley Guide to Bird Life and Behavior*, 45–46; Keffer, *Earth Almanac*, 166.
60 Biology Dictionary, "Divergent Evolution."
61 Lanner, *Made for Each Other*, 33-37; Sibley, *Sibley Guide to Bird Life and Behavior*, 411.
62 Sibley, *Sibley Guide to Bird Life and Behavior*, 411.
63 Lanner, *Made for Each Other*, 33–37.
64 Lanner, *Made for Each Other*, 30; National Park Service, "Colorado Piñon."
65 Lanner, *Made for Each Other*, 36.
66 Fitzgerald, Meaney, and Armstrong, *Mammals of Colorado*, 241, 243.
67 Mayo Clinic, "Hantavirus Pulmonary Syndrome."
68 USDA, "Poison Hemlock (*Conium maculatum*);" Whitson, *Weeds of the West*, 22–23.
69 Sibley, *Sibley Guide to Bird Life and Behavior*, 173.

70 Sibley, *Sibley Guide to Bird Life and Behavior*, 228; Audubon, "Peregrine Falcon."

71 Sibley, *Sibley Guide to Bird Life and Behavior*, 228–229.

72 Xia, "LA's Coast Was Once."

73 Keffer, *Earth Almanac*, 2020.

74 Blanton, "Torpor and Other Physiological Adaptations."

75 Kirkpatrick, *Wildflowers of the Western Plains*, 116–117.

76 Keffer, *Earth Almanac*, 78.

77 University of Colorado Museum of Natural History, "Mourning Cloak Butterfly."

78 Hammerson, *Amphibians and Reptiles in Colorado*, 336–344; Degenhardt, Painter, and Price, *Amphibians and Reptiles of New Mexico*, 295–297.

79 Degenhardt, Painter, and Price, *Amphibians and Reptiles of New Mexico*, 352.

80 Hammerson, *Amphibians and Reptiles in Colorado*, 145.

81 Scheele et al., "Amphibian Fungal Panzootic Causes," 1459–1463.

82 Hammerson, *Amphibians and Reptiles in Colorado*, 147.

83 Hammerson, *Amphibians and Reptiles in Colorado*, 139–140.

84 Pough et al., *Herpetology*, 387, 390–391; Carlson, "American Bullfrog."

85 Hammerson, *Amphibians and Reptiles in Colorado*, 125–126; Degenhardt, Painter, and Price, *Amphibians and Reptiles of New Mexico*, 72.

86 Storey and Storey, "Molecular Physiology of Freeze Tolerance," 623–665.

87 Ligon and Burt, "Evolutionary Origins," 5–34.

88 Sibley, *Sibley Guide to Bird Life and Behavior*, 477 and 563–564.

89 Sibley, *Sibley Guide to Bird Life and Behavior*, 477.

90 Fitzgerald, Meaney, and Armstrong, *Mammals of Colorado*, 317–321.

91 Whitson, *Weeds of the West*, 430–431.

92 Owen, Sieg, Johnson, and Gehring, "Exotic Cheatgrass and Loss of Soil," 2503–2517.

93 Main, "Bumblebees Are Going Extinct."

94 Main, "Bumblebees Are Going Extinct."
95 Duvuvuei. "Wildlife Corridors."
96 Angier, "Nature's Drone, Deadly and Pretty."
97 May, "Critical Overview of Progress;" Daley, "Dragonflies Embark on an Epic."
98 Stephens, "Reign of the Giant Insects."
99 Peek, "Wapiti (*Cervus elephus*)," 877–888.
100 Fitzgerald, Meaney, and Armstrong, *Mammals of Colorado*, 317–321.
101 Kirkpatrick, *Wildflowers of the Western Plains*, 1992.
102 Hammerson, *Amphibians and Reptiles in Colorado*, 215–216.
103 Degenhardt, Painter, and Price, Amphibians and Reptiles of New Mexico, 62.
104 Sibley, *Sibley Guide to Bird Life and Behavior*, 364.
105 Sibley, *Sibley Guide to Bird Life and Behavior*, 358–359.
106 Fitzgerald, Meaney, and Armstrong, *Mammals of Colorado*, 317–321.
107 Kirkpatrick, *Wildflowers of the Western Plains*, 160.
108 Muthersbaugh et al., "Maternal Behaviors Influence Survival," 2.
109 Byers, *Built for Speed*, 26–28 and 201–202.
110 Byers, *Built for Speed*, 6, 57, and 587.
111 Byers, *Built for Speed*, 17–18.
112 Deboot, Fisher, Buckhouse, and Swanson, "Monitoring Hydrological Changes," 227–232; Foxx and Tierney, "Rooting Patterns," 65–68; Tennesen, "When Juniper and Woody Plants Invade," 1630–1631.
113 St. John, "Canada Geese Can't Fly in July."
114 Mackie, Kie, Pac, and Hamlin, "Mule Deer (*Odocoileus hemionus*)," 892–893.
115 Miller, Ceballos, and Reading, "The Prairie Dog and Biotic Diversity," 677–681.
116 Fitzgerald, Meaney, and Armstrong, *Mammals of Colorado*, 302–303.
117 Fitzgerald, Meaney, and Armstrong, *Mammals of Colorado*, 302–303.
118 Pelton, "Black Bear (*Ursus americanus*)," 547.
119 Eaton and Kaufman, *Kaufman Field Guide to Insects*, 32.

120 Fitzgerald, Meaney, and Armstrong, *Mammals of Colorado*, 134.
121 Kirkpatrick, *Wildflowers of the Western Plains*, 94; Reyes, "Benefits of Navajo Tea."
122 Eaton and Kaufman, *Kaufman Field Guide to Insects*, 88–90.
123 Duvuvuei, "Wildlife Corridors: Long-billed Curlews."
124 Agrawal, *Monarchs and Milkweed*, 2017.
125 Agrawal, *Monarchs and Milkweed*, 2017.
126 Agrawal, *Monarchs and Milkweed*, 2017.
127 Agrawal, *Monarchs and Milkweed*, 2017.
128 O'Dowd and Hagan, "Why Avocados Attract;" Suárez and Blanquet, "That Super Bowl Guacamole."
129 Agrawal, *Monarchs and Milkweed*, 2017.
130 Fitzgerald, Meaney, and Armstrong, *Mammals of Colorado*, 350–351.
131 Eaton and Kaufman, *Kaufman Field Guide to Insects*, 202.
132 Peek, "Wapiti (*Cervus elephus*)," 878 and 891.
133 Cook, *Border and the Buffalo*, All Gone, 113.
134 Flores, "Long Love Affair with an Uncommon Country," 230–232.
135 Flores, "Long Love Affair with an Uncommon Country," 230–232.
136 Isenberg, *Destruction of the Bison*, 128–129.
137 Phippen, "Kill Every Buffalo."
138 Isenberg, *Destruction of the Bison*, 136.
139 Brown, *My Heart at Wounded Knee*, 1970.
140 Geremia et al., "Migrating Bison Engineer the Green Wave."
141 Frank, McNaughton, and Tracy, "The Ecology of the Earth's Grazing Ecosystems," 513–521.
142 Biggins and Kosoy, "Influences of Introduced Plague," 906–916.
143 US Fish and Wildlife Service, *Black-Footed Ferret Five-Year Status Review*, 2008.
144 Soulé and Terborgh, *Continental Conservation*, 2005; Miller et al., "Prairie Dogs: An Ecological Review," 2801–2810.
145 Uresk and Paulson, "Estimated Carrying Capacity," 387–390; Miller et al., "Prairie Dogs: An Ecological Review," 2801–2810.

146 Miller et al., "Prairie Dogs: An Ecological Review and Current Biopolitics," 2801–2810.

147 Miller et al., "Prairie Dogs: An Ecological Review and Current Biopolitics," 2801–2810.

148 Miller, Ceballos, and Reading, "Prairie Dog and Biotic Diversity," 677–681; Miller et al., "Prairie Dogs: An Ecological Review and Current Biopolitics," 2801–2810; Kotliar, Miller, Reading, and Clark, "Prairie Dog as a Keystone Species," 65–88.

149 Soulé and Terborgh, *Continental Conservation*, 2005.

150 Seamans, Rau, and Sanders, *Mourning Dove Population Status*, 1, 10, 12–14.

151 Meyer, "Assessing Mourning Dove Population Declines," 9 and 20.

152 Meyer, "Assessing Mourning Dove Population Declines."

153 Cornell Lab of Ornithology, "Wilson's Warbler Life History."

154 Hay, "Introducing the Swainson's Hawk."

155 Lewis, "Tarantulas Astir for Mating Season."

156 Harlow, "Torpor and Other Physiological Adaptations," 267–269.

157 Terborgh and Estes, *Trophic Cascades*, 2010.

158 Ripple et al., "A Shifting Ecological Baseline," 430–434.

159 Craighead. *For Everything There Is a Season*, 1994.

Reference Material

Aby, Scott B. "Date of Arroyo Cutting in the American Southwest and the Influence of Human Activities." *Anthropocene* 18 (June 2017): 76–88. https://doi.org/10.1016/j.ancene.2017.05.005.

Agrawal, Anurag. *Monarchs and Milkweed: A Migrating Butterfly, a Poisonous Plant, and Their Remarkable Story of Coevolution*. Princeton, NJ: Princeton University Press, 2017.

Angier, Natalie. "Nature's Drone, Deadly and Pretty." *The New York Times*, April 2, 2013. https://www.nytimes.com/2013/04/02/science/dragonflies-natures-deadly-drone-but-prettier.html.

Audubon. "Peregrine Falcon." Last accessed November 20, 2024. https://www.audubon.org/field-guide/bird/peregrine-falcon.

Baker, Bruce W., and Edward P. Hill. "Beaver (*Castor canadensis*)." In *Wild Mammals of North America: Biology, Management, and Conservation*. 2nd edition. Edited by George A. Feldhamer, Bruce C. Thompson, and Joseph A. Chapman, 288–310. Baltimore: John Hopkins Press, 2003.

Biggins, Dean E., and Michael Y. Kosoy. "Influences of Introduced Plague on North American Mammals: Implications from Ecology of Plague in Asia." *Journal of Mammalogy* 82, no. 4 (November 2001): 906–916. https://www.jstor.org/stable/1383469.

Biology Dictionary. "Divergent Evolution." Last updated April 28, 2017. https://biologydictionary.net/divergent-evolution/.

Black, Riley. "Why Bats Are One of Evolution's Greatest Puzzles." *Smithsonian Magazine*, April 21, 2021. https://www.smithsonianmag.com/science-nature/bats-evolution-history-180974610/.

Blanton, Kayla. "Torpor and Other Physiological Adaptations of the Badger (*Taxidae taxus*) to Cold Environments." Everyday Health (website), September 20, 2023. https://www.everydayhealth.com/brown-fat-guide/.

Block, William M., and Deborah M. Finch. "Songbird Ecology in Southwestern Ponderosa Pine Forests: A Literature Review." *General Technical Report RM-GTR-292*. Fort Collins, CO: US Department of Agriculture, Forest Service, Rocky Mountain Forest and Research Station, 1997.

Branch, Carrie L., Georgy A. Semenov, Dominique N. Wagner, Benjamin R. Sonnenburg, Angela M. Pitera, Eli S. Bridge, Scott A. Taylor, and Vladimir V. Pravosudov. "The Genetic Basis of Spatial Cognitive Variation in a Food-Caching Bird." *Current Biology* 32, R 37-R39 (January 2022): 210–219. https://doi.org/10.1016/j.cub.2021.10.036.

Brown, Dee. *Bury My Heart at Wounded Knee*. New York: Holt, Rinehart, and Winston, 1970.

Byers, John A. *Built for Speed: A Year in the Life of Pronghorn*. Cambridge, MA: Harvard University Press, 2003.

Carlson, Constance. "The American Bullfrog." Wetlands Parks Friends (website), August 14, 2022. https://wetlandsparkfriends.org/the-american-bullfrog/.

Chronic, Halka. *Roadside Geology of New Mexico*. Missoula, MT: Mountain Press Publishing, 1987.

Cook, John R. *The Border and the Buffalo, All Gone*. Austin, TX: State House Press, 1989.

Cornell Lab of Ornithology. "Wilson's Warbler Life History." All About Birds (website). Last accessed April 18, 2024. https://www.allaboutbirds.org/guide/Wilsons_Warbler/lifehistory.

Cotter, Holland. "Eugene V. Thaw, Influential Art Collector and Dealer, Is Dead at 90." *The New York Times*, January 5, 2018. https://www.nytimes.com/2018/01/05/obituaries/eugene-v-thaw-dies-art-collector-and-dealer.html?smid=url-share.

Craighead, Frank. *For Everything There Is a Season: The Sequence of Natural Events in the Grand Teton-Yellowstone Area*. Helena, MT: Falcon Press, 1994.

Daley, Jason. "Dragonflies Embark on an Epic, Multigenerational Migration Each Year: Monarch Butterflies Aren't the Only Migratory Marathoners in North America." *Smithsonian Magazine*, January 9, 2019. https://www.smithsonianmag.com/smart-news/dragonfly-undertakes-epic-multi-generational-migration-each-year-180971190/.

Deboodt, Tim L., Michael P. Fisher, J. C. Buckhouse, and John Swanson. "Monitoring Hydrological Changes Related to Western Juniper Removal: A Paired Watershed Approach." In *Proceedings of the Third Interagency Conference on Research in the Watersheds*, (September 8, 2008): 8–11.

Degenhardt, William G., Charles W. Painter, and Andrew H. Price. *Amphibians and Reptiles of New Mexico*. Albuquerque, NM: University of New Mexico Press, 1996.

Dick-Peddie, William A. *New Mexico Vegetation: Past, Present, and Future*. Albuquerque, NM: University of New Mexico Press, 1993.

Duvuvuei, Erin. "Wildlife Corridors: Long-billed Curlews." *New Mexico Wildlife* 62, no. 2 (Winter 2020). https://magazine.wildlife.state.nm.us/wildlife-pathways-long-billed-curlews/.

Eaton, Eric R., and Kenn Kaufman. *Kaufman Field Guide to Insects of North America*. Tucson, AZ: Hillstar Editions, 2007.

Fitzgerald, James P., Carron A. Meaney, and David M. Armstrong. *Mammals of Colorado*. Boulder, CO: University of Colorado Press, 1994.

Flannery, Tim. *The Eternal Frontier: An Ecological History of North America and Its Peoples*. New York: Grove Press, 2001.

Flannery, Tim. *The Future Eaters: An Ecological History of the Australasian Lands and People*. New York: Grove Press, 1994.

Fletcher, Reggie, and Wayne A. Robbie. "Historic and Current Conditions of Southwestern Grasslands." In *Assessment of Grassland Ecosystem Conditions in the Southwestern United States*. Edited by Deborah M. Finch, 120–129. USDA Forest Service General Technical Report, RMRS-GTR-135-vol. 1, 2004.

Flores, Dan. "A Long Love Affair with an Uncommon Country: Environmental History and the Great Plains." In *Prairie Conservation*. Edited by Fred B. Sampson and Fritz L Knopf, 3–18. Washington, DC: Island Press, 1996.

Flores, Dan. *Wild New World*. New York: Norton Press, 2022.

Foreman, David. *Rewilding North America: A Vision for Conservation in the 21st Century*. Washington, DC: Island Press, 2004.

Fowler, Charles W., and Larry Hobbs. "Is Humanity Sustainable?" *Proceedings of the Royal Society*, 270, no. 1533 (December 2003): 2579–2583. https://doi.org/10.1098/rspb.2003.2553.

Foxx, Teralene S., and Gail D. Tierney. "Rooting Patterns in the Pinyon-Juniper Woodland." In *Proceedings of the Pinyon-Juniper Conference*. Edited by Richard L. Everett, 65–68. Reno, NV: US Department of Agriculture, Forest Intermountain Research Station, 1987.

Frank, Douglas A., Samuel J. McNaughton, and Benjamin F. Tracy. "The Ecology of the Earth's Grazing Ecosystems: Profound Functional Similarities Exist Between the Serengeti and Yellowstone." *BioScience* 48, no. 7 (July 1998): 513–521.

Franke, Mary Anne. *To Save the Wild Bison: Life on the Edge in Yellowstone*. Norman, OK: University of Oklahoma Press, 2005.

Geremia, Chris, Jerod A. Merkle, Daniel R. Eacker, Rick L. Wallena, P. J. White, Mark Hebblewhite, and Matthew J. Kauffman. "Migrating Bison Engineer the Green Wave." *Proceedings of National Academy of Sciences* 116, no. 51 (November 2019): 25707–25713. https://doi.org/10.1073/pnas.1913783116.

Hammerson, Geoffrey A. *Amphibians and Reptiles in Colorado*. Boulder, CO: University of Colorado Press, 1991.

Harlow, Henry J. "Torpor and Other Physiological Adaptations of the Badger (*Taxidae taxus*) to Cold Environments." *Physiological and Biological Zoology* 54, no. 3 (July 1981): 267–269. https://doi.org/10.1086/physzool.54.3.30159941.

Hawk Mountain Global Raptor Conservation. "Golden Eagle." Hawk Mountain Sanctuary (website). Last accessed April 18, 2024. https://www.hawkmountain.org/raptors/golden-eagle.

Hay, Anne. "Introducing the Swainson's Hawk." Buffalo Bill Center of the West (website), October 1, 2018. https://centerofthewest.org/2018/10/01/introducing-swainsons-hawk/.

Hewitt, Godfrey. "Ice Ages: Species Distributions, and Evolution." In *Evolution on Planet Earth*. Edited by Lynn J. Rothschild and Adrian M. Lister, 339–361. San Diego: Academic Press, 2003. https://doi.org/10.1016/B978-012598655-7/50045-8.

Isenberg, Andrew C. *The Destruction of the Bison*. New York: Cambridge University Press, 2000.

Ivey, Robert DeWitt. *Flowering Plants of New Mexico*. 5th edition. Albuquerque, NM: Fiscal Book, 2008.

Keffer, Ken. *Earth Almanac: Nature's Calendar for Year-Round Discovery*. Seattle: Skipstone Press, 2020.

Kirkpatrick, Zoe M. *Wildflowers of the Western Plains*. Lincoln, NE: University of Nebraska Press, 1992.

Kotliar, Natasha B., Brian J. Miller, Richard P. Reading, and Timothy W. Clark. "The Prairie Dog as a Keystone Species." In *Conservation of the Black-Tailed Prairie Dog: Saving North America's Western Grasslands*. Edited by John L. Hoogland, 65–88. Washington, DC: Island Press, 2006.

Lanner, Ronald M. *Made for Each Other: A Symbiosis of Birds and Pines*. New York: Oxford University Press, 1996.

Leopold, Aldo. *A Sand County Almanac with Essays on Conservation from Round River*. New York: Ballantine Books, 1966.

Lewis, Olivia. "Tarantulas Astir for Mating Season in Northern New Mexico." *Santa Fe New Mexican*, September 27, 2005.

Ligon, J. David, and D. Brent Burt. "Evolutionary Origins." In *Ecology and Evolution of Cooperative Breeding in Birds*. Edited by Walter D. Koenig and Janis L. Dickinson, 5–34. Cambridge: Cambridge University Press, 2004.

Lott, Dale F. *American Bison: A Natural History*. Berkeley: University of California Press, 2002.

Mac, Michael J., Paul A. Opler, Catherine E. Puchett Haecker, and Peter D. Doran. *Status and Trends of the Nation's Biological Resources*. Reston, VA: US Department of the Interior, US Geological Survey, 1998.

MacArthur, R. H., and E. O. Wilson. *The Theory of Island Biogeography*. Princeton, NJ: Princeton University Press, 1967.

Mackie, Richard J., John G. Kie, David F. Pac, and Kenneth L. Hamlin. "Mule Deer (*Odocoileus hemionus*)." In *Wild Mammals of North America: Biology, Management, and Conservation*. 2nd edition. Edited by George A. Feldhamer, Bruce C. Thompson, and Joseph A. Chapman, 889–905. Baltimore: John Hopkins Press, 2003.

Main, Douglas. "Bumblebees Are Going Extinct in a Time of 'Climate Chaos.'" *National Geographic*, February 6, 2020. https://www.nationalgeographic.com/animals/2020/02/bumblebees-going-extinct-climate-change-pesticides/.

Martin, Paul S. *Twilight of the Mammoths: Ice Age Extinctions and the Rewilding of America*. Berkeley: University of California Press, 2007.

Martin, Paul S., and Richard G. Klein, eds. *Quaternary Extinctions: A Prehistoric Evolution*. Tucson, AZ: University of Arizona Press, 1984.

Mason, Betsy. "Total Recall: A Brilliant Memory Helps Chickadees Survive." *Knowable Magazine*, September 5, 2019. https://knowablemagazine.org/article/mind/2019/chickadee-memory-food.

Matthiessen, Peter. *Wildlife in America*. New York: Viking, 1987.

May, Michael L. "A Critical Overview of Progress in Studies of Dragonflies (ODONATA: *Anisoptera*), with Emphasis on North America." *Journal of Insect Conservation* 17 (November 2012): 1–15. https://link.springer.com/article/10.1007/s10841-012-9540-x.

Mayo Clinic. "Hantavirus Pulmonary Syndrome." Mayo Clinic (website). Accessed November 1, 2023. https://www.mayoclinic.org/diseases-conditions/hantavirus-pulmonary-syndrome/symptoms-causes/syc-20351838.

Meyer, Paul M. *Assessing Mourning Dove Population Declines: Changes in Nesting Dynamics and the Role of Perch Sites*. MS thesis, Utah State University, 1994. https://digitalcommons.usu.edu/etd/6499/.

Miller, Brian J., Henry J. Harlow, Tyler Harlow, Dean Biggins, and William J. Ripple. "Trophic Cascades Linking Wolves (*Canis lupus*), Coyotes (*Canis latrans*), and Small Mammals." *Canadian Journal of Zoology* 90, no. 1 (January 2012): 70–78. https://doi.org/10.1139/z11-115.

Miller, Brian J., Richard P. Reading, Dean Biggins, James Detling, Steve Forrest, John Hoogland, Sterling Miller, Jonathan Proctor, Joe Truett, Jody Javersak, and Daniel W. Uresk. "Prairie Dogs: An Ecological Review and Current Biopolitics." *Journal of Wildlife Management* 71, no. 8 (November 2007): 2801–2810. http://www.jstor.org/stable/4496405.

Miller, Brian, Gerardo Ceballos, and Richard Reading. "The Prairie Dog and Biotic Diversity." *Conservation Biology* 8, no. 3 (September 1994): 677–681. https://www.jstor.org/stable/2386509.

Muthersbaugh, Michael S., Wesley W. Boone, Elizabeth A. Saldo, Alex J. Jensen, Jay Cantrell, Charles Ruth, John C. Kilgo, and David S. Jachowski. "Maternal Behaviors Influence Survival of Ungulate Neonates Under Heavy Predation Risk." *Ecology and Evolution* 14, no. 8 (August 2024): e70151. https://doi.org/10.1002/ece3.70151.

National Park Service. "Colorado Piñon." Bryce Canyon. National Park Service (website). Last updated September 4, 2021. https://www.nps.gov/brca/learn/nature/pinyonpine.htm.

New Mexico Department of Game and Fish. *Comprehensive Wildlife Conservation Strategy for New Mexico*. Sante Fe, NM: New Mexico Department of Game and Fish, 2006.

New Mexico State University. "Loamy Upland." Ecosystem Dynamics Interpretive Tool (website). Last accessed November 20, 2024. https://edit.jornada.nmsu.edu/catalogs/esd/070A/R070AY001NM.

O'Dowd, Peter, and Allison Hagan. "Why Avocados Attract the Interest of Mexican Drug Cartels." WBUR National Public Radio, February 7, 2020. https://www.wbur.org/hereandnow/2020/02/07/avocados-mexican-drug-cartels.

Ortiz, Alphonso. "Popay's Leadership: A Pueblo Perspective on the 1680 Revolt." In *Telling New Mexico: A New History*. Edited by Marta Weigle, 107–114. Santa Fe, NM: Museum of New Mexico Press, 2009.

Owen, Susan M., Carolyn Hull Sieg, Nancy Collins Johnson, and Catherine A. Gehring. "Exotic Cheatgrass and Loss of Soil Biota Decrease the Performance of a Native Grass." *Biological Invasions* 15, no. 9 (September 2013): 2503–2517. https://doi.org/10.1007/s10530-013-0469-0.

Paige, Christine. *Alberta Landholder's Guide to Wildlife Friendly Fencing*. Sherwood Park: Alberta Conservation Association, 2020.

Peek, James M. "Wapiti (*Cervus elephus*)." In *Wild Mammals of North America: Biology, Management, and Conservation*. 2nd edition. Edited by George A. Feldhamer, Bruce C. Thompson, and Joseph A. Chapman, 877–888. Baltimore: John Hopkins Press, 2003.

Pelton, Michael R. "Black Bear (*Ursus americanus*)." In *Wild Mammals of North America: Biology, Management, and Conservation*. 2nd edition. Edited by George A. Feldhamer, Bruce C. Thompson, and Joseph A. Chapman, 547–555. Baltimore: John Hopkins Press, 2003.

Phippen, J. Weston. "Kill Every Buffalo You Can. Every Buffalo Dead Is an Indian Gone." *The Atlantic,* May 16, 2016. https://www.theatlantic.com/national/archive/2016/05/the-buffalo-killers/482349/.

Pough, F. Harvey, Robin M. Andrews, Martha L. Crump, Alan H. Savitzky, Kentwood D. Wells, and Mathew C. Brandley. *Herpetology.* 4th edition. Sunderland, MA: Sinauer Associates. 2018.

Price, L. Greer, ed. *The Geology of Northern New Mexico's Parks, Monuments, and Public Lands.* Socorro: New Mexico Bureau of Geology and Minerals, 2010.

Reading, Richard P., Steven R. Beissinger, John J. Grensten, and Tim W. Clark. "Attributes of Black-Tailed Prairie Dog Colonies in Northcentral Montana, with Management Recommendation for Conservation of Biodiversity." *Montana Bureau of Land Management Wildlife Technical Bulletin* 2 (January 1989): 13–27.

Reyes, Bianca. "Benefits of Navajo Tea." Medium (website), October 21, 2021. https://biancawrotethis.medium.com/benefits-of-navajo-tea-811c961807f4.

Ripple, William J., Christopher Wolf, Robert L. Beschta, Apryle D. Craig, Zachary S. Curcija, Erick J. Lundgren, Lauren C. Satterfield, Samuel T. Woodrich, and Aaron J. Wirsing. "A Shifting Ecological Baseline After Wolf Extirpation." *BioScience* 74, no. 7 (July 2024): 430–434. https://doi.org/10.1093/biosci/biae034.

Scheele, Ben C., Frank Pasmans, Lee F. Skerratt, Lee Berger, An Martel, Wouter Beukema, Aldemar A. Acevedo et al. "Amphibian Fungal Panzootic Causes Catastrophic and Ongoing Losses of Biodiversity." *Science* 363, no. 6434 (March 2019): 1459–1463. https://doi.org/10.1126/science.aav0379.

Seamans, Mark E., Rebecca D. Rau, and Todd A. Sanders. *Mourning Dove Population Status, 2013.* Washington, DC: US Department of the Interior, Fish and Wildlife Service, Division of Migratory Bird Management, 2013.

Sibley, David A. *Sibley Guide to Bird Life and Behavior.* New York: Alfred A. Knopf, 2001.

Smith, Anna V. "The State of Tribal Co-Management of Public Lands." *High Country News,* September 22, 2023. https://www.hcn.org/articles/public-lands-the-state-of-tribal-co-management-of-public-lands.

Smith, Robert B., and Lee J. Siegel. *Windows into the Earth: The Geologic Story of Yellowstone and Grand Teton National Parks*. New York: Oxford University Press, 2000.

Soulé, Michael E., and John Terborgh, eds. *Continental Conservation: Scientific Foundations of Regional Reserve Networks*. Covelo, CA: Island Press, 1999.

Speakman, John R. "The Evolution of Flight and Echolocation in Bats: Another Leap in the Dark." *Mammal Review* 31, no. 2 (December 2001): 111–130. https://doi.org/10.1046/j.1365-2907.2001.00082.x.

St. John, Kate. "Canada Geese Can't Fly in July." *Outside My Window: A Blog of Birds and Nature*, June 14, 2023. https://www.birdsoutsidemywindow.org/2023/06/14/canada-geese-cant-fly-in-july/.

Stephens, Tim. "Reign of the Giant Insects Ended with the Evolution of Birds." *UC Santa Cruz Newsletter*, June 4, 2012. https://news.ucsc.edu/2012/06/giant-insects.html.

Storey, Kenneth B., and Janice M. Storey. "Molecular Physiology of Freeze Tolerance in Vertebrates." *Physiological Review* 97, no. 2 (April 2017): 623–665. https://doi.org/10.1152/physrcv.00016.2016.

Suárez, Karol, and Christopher Rogel Blanquet. "That Super Bowl Guacamole You're Eating Probably Had a Risky Trip Through Cartel Territory." *Courier Journal*, February 8, 2023. https://www.courier-journal.com/story/news/crime/2023/02/08/rival-drug-cartels-profit-mexican-avocados-us-police-work/69840227007/.

Tennesen, Michael. "When Juniper and Woody Plants Invade, Water May Retreat." *Science* 322, no. 5908 (December 2008): 1630–1631. https://doi.org/10.1126/science.322.5908.1630.

Terborgh, John, and James E. Estes, eds. *Trophic Cascades: Predators, Prey, and the Changing Dynamics of Nature*. Washington, DC: Island Press, 2010.

University of Colorado Museum of Natural History. "Mourning Cloak Butterfly." Museum of Natural History Research and Innovation Office (website). Last updated August 25, 2021. https://www.colorado.edu/cumuseum/2021/08/25/mourning-cloak-butterfly.

Uresk, Daniel W., and Deborah B. Paulson. "Estimated Carrying Capacity for Cattle Competing with Prairie Dogs and Forage Utilization in Western South Dakota." In *Symposium on Management of Amphibians, Reptiles, and Small Animals in North America*, 387–390. Washington, DC: USDA Forest Service General Technical Report RM-166, 1989.

US Fish and Wildlife Service. *Birds of Conservation Concern.* Arlington, VA: US Fish and Wildlife Service Division of Migratory Bird Management, 2008.

US Fish and Wildlife Service. *Black-Footed Ferret Five-Year Status Review: Summary and Evaluation.* Pierre, SD: US Fish and Wildlife Service South Dakota Field Office, 2008.

US Fish and Wildlife Service. *Rio Mora National Wildlife Refuge and Conservation Area: A Land Protection Plan.* Albuquerque, NM: US Fish and Wildlife Service National Wildlife Refuge System Southwest Region Division of Planning, 2012.

USDA. "Poison Hemlock (*Conium maculatum*)." USDA Agriculture Research Service (website). Last updated June 26, 2018. https://www.ars.usda.gov/pacific-west-area/logan-ut/poisonous-plant-research/docs/poison-hemlock-conium-maculatum/.

Ward, Peter D. *The Call of the Distant Mammoths: Why the Ice Age Mammals Disappeared.* New York: Copernicus, Springer-Verlag, 1997.

Whitson, Tom D., ed. *Weeds of the West.* Jackson, WY: Western Society of Weed Science in cooperation with the Western United States Land Grant Universities Cooperative Extension Services, 2006.

Xia, Rosanna. "LA's Coast Was Once a DDT Dumping Ground: No One Could See It—Until Now." *Los Angeles Times*, October 25, 2020. https://www.latimes.com/projects/la-coast-ddt-dumping-ground/.

Zeedyk, Bill, and Jan-Willem Jansens. *An Introduction to Erosion Control.* Santa Fe, NM: Earth Works Institute and Quivira Coalition, 2009.

Zeedyk, Bill, and Van Clothier. *Let the Water Do the Work: Induced Meandering, an Evolving Method for Restoring Incised Channels.* Santa Fe, NM: Quivira Coalition, 2009.

Zeedyk, Bill, Mollie Walton, and Tamara Gadzia. *Characterization and Restoration of Slope Wetlands in New Mexico: A Guide for Understanding Slope Wetlands, Causes of Degradation, and Treatment Options.* Santa Fe, NM: New Mexico Environment Department Surface Water Quality Bureau, Wetlands Program, 2014.

Zeedyk, Bill. *An Introduction to Induced Meandering: A Method of Restoring Stability to Incised Stream Channels.* Santa Fe, NM: Paper Tiger, 2009.

Zeedyk, Bill. *Water Harvesting from Low Standard Rural Roads.* Santa Fe: New Mexico Environment Department, 2012.

About the Author

Brian Miller received his PhD from the University of Wyoming in 1988 with a focus on behavioral ecology and conservation of the endangered black-footed ferret. In 1989, he was awarded a Smithsonian Institution Fellowship at the Conservation and Research Center of the National Zoological Park centered on preparing captive-raised black-footed ferrets for reintroduction onto the western prairies.

From 1992 to 1997, Miller lived and worked in Mexico as a post-doc and then professor at the National University of Mexico. He was part of a team starting a protected area in the high plains of Chihuahua, Mexico, before beginning an ongoing research project on jaguars and pumas in the dry tropical forest of Jalisco, Mexico. In 1997, Miller and Rich Reading started the Conservation Department at the Denver Zoological Foundation. While there, Miller examined the trophic impacts of wolf reintroduction on coyotes and the small mammal community in Grand Teton National Park.

Miller's main research interest concerns the role of highly interactive species in regulating ecosystem processes and how to improve protection for those species when designing reserves. In 2005, he became executive director and founding scientist of the Wind River Ranch Foundation, an NGO located on a private ranch owned by Eugene and Clare Thaw and dedicated to conservation of native species, conservation research, and environmental education. In 2012, the land officially entered the US Fish and Wildlife Service (USFWS) as the Rio Mora National Wildlife Refuge.

Miller has co-authored or edited five books and has published over one hundred scientific articles. He has served on the board of directors for five NGOs and as a scientific advisor for several more. He has received outstanding service awards from the Colorado Division of Fish and Wildlife for co-heading the Lynx Advisory team and from the USFWS for his work on black-footed ferret conservation. In 2009, he received the Conservationist of the Year Award from the Denver Zoological Foundation.

About BRG Scientific

BRG Scientific was founded to advance scholarly works of the highest intellectual quality. We place priority on books showcasing the topics of ecology, natural history, evolution, animal behavior, and comparative biology written by vetted colleagues. Our team is driven by excellence of scholarship, not by volume of sales. We strive to reach technical audiences as well as curious, non-technical readers wishing to expand the horizons of their knowledge. We recognize that rigorous, well-researched science can also be pleasing to the eye and soul and that beautiful and informative graphic illustrations complement and enhance the written narrative.

See more from this publisher at:
www.BRGScientific.com

www.ingramcontent.com/pod-product-compliance
Lightning Source LLC
Chambersburg PA
CBHW040934030426
42337CB00006B/55